Human Motion Simulation

The first author dedicates this book to his parents Josette and Tony, his wife Mary, and his children Ella, Tony, and Sam

The second author dedicates the book to his wife Rita and his daughter Ruhee, and to the memory of his parents

Human Motion Simulation
Predictive Dynamics

Karim A. Abdel-Malek

Jasbir Singh Arora

The Virtual Soldier Research Program
The University of Iowa

AMSTERDAM • BOSTON • HEIDELBERG • LONDON
NEW YORK • OXFORD • PARIS • SAN DIEGO
SAN FRANCISCO • SINGAPORE • SYDNEY • TOKYO
Academic Press is an imprint of Elsevier

Academic Press is an imprint of Elsevier
225 Wyman Street, Waltham, MA 02451, USA
525 B Street, Suite 1800, San Diego, CA 92101-4495, USA
32 Jamestown Road, London NW1 7BY, UK

Notice

Knowledge and best practice in this field are constantly changing. As new research and
experience broaden our understanding, changes in research methods, professional
practices, or medical treatment may become necessary.

Practitioners and researchers must always rely on their own experience and knowledge in
evaluating and using any information, methods, compounds, or experiments described
herein. In using such information or methods they should be mindful of their own safety
and the safety of others, including parties for whom they have a professional the fullest
extent of the law, neither the Publisher nor the authors, contributors, or editors, assume any
liability for any injury and/or damage to persons or property as a matter of products liability, negligence or otherwise, or from any use or operation of any methods, products,
instructions, or ideas contained in the material herein.

British Library Cataloguing-in-Publication Data
A catalogue record for this book is available from the British Library

Library of Congress Cataloging-in-Publication Data
A catalog record for this book is available from the Library of Congress

ISBN: 978-0-12-405190-4

For information on all Academic Press publications
visit our website at elsevierdirect.com

Typeset by MPS Limited, Chennai, India
www.adi-mps.com

Printed and bound in the United States of America

13 14 15 16   10 9 8 7 6 5 4 3 2 1

Contents

Preface xiii
Acknowledgments xv

CHAPTER 1 Introduction..**1**
    1.1 What is predictive dynamics? 1
    1.2 How does predictive dynamics work? 2
    1.3 Why data-driven human motion prediction does not work 3
    1.4 Concluding remarks 4
    References 5

CHAPTER 2 Human Modeling: Kinematics....................................**7**
    2.1 Introduction 7
    2.2 General rigid body displacement 10
           2.2.1 Example: rotation and translation 11
    2.3 Concept of extended vectors and homogeneous coordinates......13
    2.4 Basic transformations 14
           2.4.1 Example: knee rotation 16
    2.5 Composite transformations 17
           2.5.1 Example: composite transformations.............................17
    2.6 Directed transformation graphs..19
           2.6.1 Example: multiple transformations................................20
    2.7 Determining the position of a multi-segmental link: forward
           kinematics...24
    2.8 The Denavit–Hartenberg representation 25
    2.9 The kinematic skeleton 27
    2.10 Establishing coordinate systems 30
           2.10.1 Example: a 9-DOF model of an upper limb...................31
           2.10.2 Example: DH parameters of the lower limb 32
    2.11 The Santos® model...36
    2.12 Variations in anthropometry 36
    2.13 A 55-DOF whole body model 37
    2.14 Global DOFs and virtual joints..39
    2.15 Concluding remarks 40
    References 40

CHAPTER 3 Posture Prediction and Optimization **41**
    3.1 What is optimization? 41
    3.2 What is posture prediction? 41

3.3   Inducing behavior...43
3.4   Posture prediction versus inverse kinematics............................44
    3.4.1   Analytical and geometric IK methods..........................44
    3.4.2   Empirically-based posture prediction 44
3.5   Optimization-based posture prediction......................................45
    3.5.1   Design variables..46
    3.5.2   Constraints 47
    3.5.3   Cost function 47
3.6   A 3-DOF arm example...47
3.7   Development of human performance measures 49
    3.7.1   Joint displacement...50
    3.7.2   Effort...50
    3.7.3   Delta potential energy 51
    3.7.4   Discomfort..53
    3.7.5   Single-objective optimization 55
    3.7.6   Numerical solutions to optimization problems.............57
3.8   Motion between two points..58
3.9   Joint profiles as B-spline curves 58
3.10 Motion prediction formulation...60
    3.10.1 Design variables 60
    3.10.2 Constraints 60
3.11 A 15-DOF motion prediction...61
    3.11.1 The 15-DOF Denavit–Hartenberg model 61
3.12 Optimization algorithm..62
3.13 Motion prediction of a 15-DOF model 63
3.14 Multi-objective problem statement...65
3.15 Design variables and constraints 65
3.16 Concluding remarks...65
References..66

CHAPTER 4   Recursive Dynamics..69
4.1   Introduction 69
4.2   General static torque 70
4.3   Dynamic equations of motion..72
4.4   Formulation of regular Lagrangian equation..............................74
    4.4.1   Sensitivity analysis..75
4.5   Recursive Lagrangian equations 75
    4.5.1   Forward recursive kinematics 76
    4.5.2   Backward recursive dynamics 76
    4.5.3   Sensitivity analysis..77

	4.5.4	Kinematics sensitivity analysis	77
	4.5.5	Dynamics sensitivity analysis	78
	4.5.6	Joint profile discretization	80
4.6		Examples using a 2-DOF arm	81
	4.6.1	The DH parameters	82
	4.6.2	Forward recursive kinematics	83
	4.6.3	Backward recursive dynamics	84
	4.6.4	Gradients	84
	4.6.5	Closed-form equations of motion	86
4.7		Trajectory planning example	87
4.8		Arm lifting motion with load example	88
4.9		Concluding remarks	90
		References	92

CHAPTER 5	**Predictive Dynamics**	**95**
5.1	Introduction	95
5.2	Problem formulation	95
5.3	Dynamic stability: zero-moment point	99
5.4	Performance measures	101
5.5	Inner optimization	102
5.6	Constraints	103
	5.6.1 Feasible set	104
	5.6.2 Minimal set of constraints	104
5.7	Types of constraints	105
	5.7.1 Time-dependent constraints	105
	5.7.2 Time-independent constraints	107
5.8	Discretization and scaling	108
5.9	Numerical example: single pendulum	109
	5.9.1 Description of the problem	109
	5.9.2 Simple swing motion with boundary conditions—PD solution	111
	5.9.3 Oscillating motion with boundary conditions—PD solution	114
	5.9.4 Oscillating motion with boundary conditions and one state-response constraint—PD solution	116
	5.9.5 Oscillating motion with boundary conditions and two state-response constraints	118
5.10	Example formulations	120
5.11	Concluding remarks	120
	References	125

CHAPTER 6   Strength and Fatigue: Experiments and Modeling......127
    6.1 Joint space 127
    6.2 Strength influences..128
    6.3 Strength assessment...132
    6.4 Normative strength data 134
    6.5 Representing strength percentiles 137
    6.6 Mapping strength to digital humans: strength surfaces............138
    6.7 Fatigue 140
    6.8 Strength and fatigue interaction....................................145
    6.9 Concluding remarks 145
    References..145

CHAPTER 7   Predicting the Biomechanics of Walking 149
    7.1 Introduction 149
    7.2 Joints as degrees of freedom (DOF)..............................151
    7.3 Muscle versus joint space 151
    7.4 Spatial kinematics model 152
        7.4.1 A kinematic 55-DOF human model 152
        7.4.2 Global DOFs and virtual joints......................................154
        7.4.3 Forward recursive kinematics 155
    7.5 Dynamics formulation..156
        7.5.1 Backward recursive dynamics 156
        7.5.2 Sensitivity analysis...157
        7.5.3 Mass and inertia property 157
    7.6 Gait model..158
        7.6.1 One-step gait model 158
        7.6.2 Ground reaction forces (GRF) 159
    7.7 Zero-Moment point (ZMP) 161
        7.7.1 Global forces at the pelvis 162
        7.7.2 Global forces at origin 163
        7.7.3 ZMP calculation 163
    7.8 Calculating ground reaction forces (GRF) 164
    7.9 Optimization formulation..166
        7.9.1 Design variables 166
        7.9.2 Objective function 166
        7.9.3 Constraints 167
    7.10 Numerical discretization 171
    7.11 Example: predicting the gait...172
        7.11.1 Normal walking..172
    7.12 Cause and effect...176

7.13 Implementations of the predictive dynamics walking
formulation 183
    7.13.1 Effect of constrained joints 183
    7.13.2 Sideways and backward walking 183
    7.13.3 Effect of changing anthropometry 183
    7.13.4 Effect of changing loads 183
    7.13.5 Walking on uneven terrains 184
    7.13.6 Asymmetric walking 184
    7.13.7 Walking on different terrain types................184
7.14 Concluding remarks 184
References 185

CHAPTER 8 Predictive Dynamics: Lifting 187
  8.1 Human skeletal model...187
  8.2 Equations of motion and sensitivities........................187
    8.2.1 Forward recursive kinematics 187
    8.2.2 Backward recursive dynamics 189
    8.2.3 Sensitivity analysis 189
  8.3 Dynamic stability and ground reaction forces (GRF).............190
  8.4 Formulation 191
    8.4.1 Lifting task 191
  8.5 Predictive dynamics optimization formulation.............192
    8.5.1 Design variables and time discretization 193
    8.5.2 Objective functions 194
    8.5.3 Constraints..194
  8.6 Computational procedure for multi-objective optimization......197
    8.6.1 Lifting determinants and error quantification...............198
  8.7 Predictive dynamics simulation 199
  8.8 Validation 201
  8.9 Concluding remarks 204
References 204

CHAPTER 9 Validation of Predictive Dynamics Tasks 207
  9.1 Introduction 207
  9.2 Motion determinants 209
  9.3 Motion capture systems..209
    9.3.1 Overview 209
    9.3.2 Optical motion capture systems....................210
    9.3.3 Marker placement protocol 211
    9.3.4 Subject preparation and data collection........212
  9.4 Methods 213

        9.4.1 Normalizing the data...213
        9.4.2 Validation methodology..214
    9.5 Validation of predictive walking task...216
        9.5.1 Walking task description..216
        9.5.2 Walking determinants 217
        9.5.3 Participants...217
        9.5.4 Results 217
    9.6 Validation of box-lifting task...224
        9.6.1 Lifting task description 224
        9.6.2 Box-lifting determinants 225
        9.6.3 Participants...225
        9.6.4 Results 225
    9.7 Feedback to the simulation 233
    9.8 Concluding remarks 233
    References..234

CHAPTER 10 Concluding Remarks...237
    10.1 Benefits of predictive dynamics 237
        10.1.1 Using the Denavit−Hartenberg (DH) method is
                effective in modeling human kinematics.......................237
        10.1.2 Predictive dynamics solves dynamics
                without integration 238
        10.1.3 Predictive dynamics renders natural motion 238
        10.1.4 Predictive dynamics induces natural behavior 238
        10.1.5 Predictive dynamics admits cause and effect...............238
        10.1.6 Predictive dynamics uses joint space,
                not muscle space 239
        10.1.7 Predictive dynamics uses dynamic strength surfaces....239
        10.1.8 The PD validation process is effective..........................240
    10.2 Applications 240
        10.2.1 Ergonomics...240
        10.2.2 Simulating an injury or a disability 240
        10.2.3 Sports biomechanics and kinesiology...........................241
        10.2.4 Human performance...241
        10.2.5 Testing equipment, digital prototyping, human
                systems integration...241
        10.2.6 Egress/ingress 242
        10.2.7 Unsafe situations 242
    10.3 Future research 243
        10.3.1 Soft-tissue dynamics 243
        10.3.2 Intelligence 243

10.3.3 Psychological and physiological factors.........................243
10.3.4 Modeling with a high level of fidelity...........................244
10.3.5 Real-time simulation...244
Reference 245

Bibliography 247
Index 269

Preface

Realistic human motion simulation is a challenging problem from both the computational as well as formulation points of view. Realistic formulation requires large degree-of-freedom mechanical and biomedical models. Usually, the models are highly redundant, posing challenges for computational schemes to produce realistic human motion. In the past, practical applications of human simulation have been formulated as statics problem (posture prediction) or simply as animated figures that provide a schematic representation of the motion, but significantly lack the physics of the motion. The need for a deep understanding of the physics of the motion on the one hand, and more importantly, the ability to obtain a cause and effect on the other hand, are critical aspects and are the subject of this book.

Indeed, the dynamics of the system must be considered in the formulation if realistic forces and torques at various joints are desired, and furthermore, to develop a strategy that leads to a true predictive formulation. We shall define the idea of "predictive dynamics" in this book as a methodology and accompanying formulation for the prediction of a physics-based motion that responds to various parameters affecting the human body.

Interaction of the human body with the external environment presents another challenge in the formulation of the problem.

The purpose of this book is to present a new approach for human motion simulation, which we call predictive dynamics. It is an optimization-based approach in which at least one or several human performance measures are optimized subject to constraints on the physics of the problem. The constraints include strength limits, joint ranges of motion, laws of physics, contact conditions with the environment, and constraints for the specific task at hand. Constraints can be as many as necessary and can include task-specific constraints. The equations of motion are treated as equality constraints in the formulation thus avoiding the cumbersome process of integrating them with some side constraints on the variables. The mere fact of not requiring the integration of the equations of motion is itself a significant advantage and has been shown to have a significant impact on this field.

Chapter 1 introduces the basic concept of predictive dynamics and explains limitations of the data-driven approaches for human motion simulation. Chapter 2 focuses on the kinematics of the human body. Various degrees of freedom (DOF) of the human body are explained and a skeletal model of the body is described. The Denavit–Hartenberg transformation approach is described as a systematic method to formulate the kinematics of the human body. Chapter 3 introduces basic concepts related to the optimization approach. The posture prediction problem is formulated as an optimization problem that is illustrated with examples.

Chapter 4 describes a formulation for the dynamics of the problem using the recursive Lagrangian approach. The kinematics and dynamics sensitivity analysis needed in the optimization approach is described. Forward recursive kinematics and backward recursive dynamics approaches are illustrated with examples. Chapter 5 contains details of the predictive dynamics approach. Dynamic stability constraint is explained. Performance measures are described. Time-dependent and time-independent constraints are explained. Discretization of the problem for numerical calculations is explained. Simple examples are included to illustrate the concepts and methods. Chapter 6 focuses on the strength and fatigue issues related to modeling of human activities. How the strength is measured and transcribed into the form that is usable in simulation of the human motion is explained.

Chapter 7 illustrates the predictive dynamics approach for the human walking problem. Formulation of the problem is described in detail and various variations of the problem (asymmetric walk, sideways walk, backward walk) are discussed. Chapter 8 describes the human lifting problem. Predictive dynamics formulation is illustrated as a multi-objective optimization problem. Chapter 9 focuses on the process of validating predictive dynamics models. The data collection and data analysis approaches are described. The determinants for a task to be validated are discussed. The validation procedure is illustrated for the human walking and lifting tasks. Finally, Chapter 10 presents some concluding remarks on benefits of the predictive dynamics approach, applications of the methodology, and thoughts on future research directions.

Acknowledgments

The authors extend their gratitude and deep appreciation to all their staff, students, and faculty colleagues who have worked on this project over the past 10 years. SANTOS®, the virtual soldier, has been a culmination of the involvement of over 200 people who have contributed in so many ways to the method, formulation, programming, ideas, and growth of the program over the years. The authors are grateful to the many funding agencies that have sponsored this work, particularly all branches of the US Military.

Karim A. Abdel-Malek (first author): "I would like to thank my friend, colleague, and 'guru' Prof. Jasbir Arora for all his wisdom, friendship, and patience throughout these past 10 years. It has been a wonderful journey and I am honored to have done it with you, Jas."

Jasbir Singh Arora (second author): "It has been a pleasure and privilege to work with Karim on a very exciting project on digital human modeling and simulation. His vision and hard work have been at the core of this very successful project that has resulted in the development of predictive dynamics concepts."

The authors wish to thank the following agencies and companies for sponsoring their research over the past 10 years:

- US Army TACOM
- US Army RDECOM
- US Navy
- US Office of Naval Research
- Naval Health Research Center
- US Marine Corps
- NAVAIR
- US Army Natick Soldier Research, Development, and Engineering Center
- Caterpillar
- John Deere
- Rockwell Collins
- United States Council for Automotive Research (Ford, General Motors, and Chrysler)
- Ford Motor Company
- Medtronic
- Korean Agency for Defense Development
- The University of Iowa
- The State of Iowa (IEAD)
- HNI Industries

The authors wish to acknowledge the following faculty colleagues, scientists, staff, and students for their instrumental and dedicated contribution to the VSR program. Santos® would not be alive today without your personal contributions...thank you!

Current Faculty and Staff

Andy Taylor
Anith Mathai
Cesar Bonezzi
Chris Murphy
Corey Goodman
H. John Yack, PhD
Harod Stauss, PhD
Hyun-Joon Chung, PhD
Jasbir Arora, PhD
Jia Lu, PhD
Karim Abdel-Malek, PhD
Kevin Kregel, PhD
Kim Farrell
Laura Frey Law, PhD
Mahdiar Hariri, PhD
Nicole Grosland, PhD
Rajan Bhatt, PhD
Rich Lineback
Salam Rahmatalla, PhD
Santanu Mukhopadhyay
Soura Dasgupta, PhD
Steve Beck
Sultan Sultan, PhD
Tim Marler, PhD
Yujiang Xiang, PhD

Current Students

Aasim Shaik
Abhinav Sharma
Anas Nassar
Angela Dani
Ben Goerdt
Ben Weintraub
Caleb Boyd
Caley Medinger
David Bein
Emmalei Huber
Haowen Xu
Jake Kersten

Jeff Rabovsky
Jingzhu Xu
Jocelyn Todd
John Looft
John Meusch
John Nicholson
Jingzhu Xu
Kaustubh Patwardhan
Kristen Spurrier
Logan Butler
Meenal Khandelwal
Michael Steiff
Mohammad Bataineh
Mollie Knake
Nic Capdevila
Prajwal Kedilaya
Richard Degenhardt III
Robert Hofer
Tom Ebnet

Past Students and Collaborators

Adam Sheer
Afroza Ali
Aidan Murphy (John Deere, Des Moines, IA)
Alex Zhou, PhD (Sr. Engineer, CFDRC)
Allison Sotckdale (Veteran's Administration, California)
Amos Patrick (CyberAnatomy, Corp, Estes Park, CO)
Andrea Laake (Nestle Purina, Inc., Davenport, IA)
Anith Mathai (Research Scientist, VSR)
Asghar Bhatti, PhD
Brent Rochambeau (CyberAnatomy, Inc.)
Brian Smith (Pearson, Iowa City, IA)
Caleb Boyd
Carl Fruehan (SantosHuman, Inc.)
Dan Anton, PhD (Eastern Washington University, Cheney, Washington)
David Janiczek
Edward Wang
Emily Horn (John Deere, Davenport, IA)
Erik Cole
Evelyn Ross (MPH, The University of Iowa)
Faisal Goussous (Rockwell Collins, Cedar Rapids, IA)

Gary Pierce (Cerner, Kansas City, Kansas)

Huda Bazyan (Information Technology Systems, University of Virginia Charlottesville)

Hyung Joo Kim, PhD (Researcher, Hyundai, Korea)

Hyung-Joon Kwon, PhD (Post doc at Ohio State University)

Hyun-Joon Chung (Post doctoral research scientist, VSR)

Imran Pirwani

Jason Olmstead (Rockwell Collins, Cedar Rapids, IA)

Jason Potratz (Square D, Cedar Rapids, IA)

Jingzheng Li (PhD Program, Statistics, The University of Iowa)

Jingzhou Yang, PhD (Assistant Professor, Texas Tech University)

Jocelyn Todd (UI)

Joo H. · Kim, PhD (Assistant Professor, Polytechnic Institute of New York University)

Jorge Luis Carmona

Jun Hyeak Choi

Katha Sheth (Iowa City, IA)

Keith Avin, PhD (Pittsburgh, PA, post doctoral scholar)

Kyle Collins (Perfect Game USA, Cedar Rapids, IA)

Lauren Graupner (PhD program University of Iowa, Iowa City, IA)

Lindsey Knake (UI Medical School)

Magnus Wu

Matt Rasmussen (HDR, Inc.)

Matt Schikore

Meagan Shanahan (Medical Murray, North Barrington, IL)

Mike Lyons (Applied Biosystems, Someiville, MA)

Molly Patrick (Denver, CO)

Naruedon Bhatarakamol (MS program, University of Pennsylvania)

Nate Horn (Red Bull Racing, Mooresville, NC)

Noah Abrahamson (Rockwell Collins, Cedar Rapids, IA)

Owen Flatley

Owen Sessions (The University of Iowa, Iowa City, IA)

Pradeep Balasundaram

Qian Wang, PhD (Assistant Professor, Manhattan College)

Rajeev Penmatsa (PhD Program, Computer Science, The University of Iowa)

Rebecca (Becca) Fetter (LMS, Coralville, IA)

Rosalind Smith (Chicago, IL)

Ross Johnson (Mazira Inc.)

Ryan Vignes (PhD program, University of Michigan; Simula Research, Jah, Norway)

Sandra Dandash, PhD (United Technologies Research Center)

Srikanth Deshpande

Stephanie Swiatlo

Tariq Sinokrot, PhD (LMS International, Coralville, IA)

Tim Marler, PhD (Research Scientist, VSR)
Ting Xia, PhD (Assistant Professor, Palmer College of Chiropractic, Davenport, IA)
Trucy Phan (MS Program, University of Berkeley)
Uday Verma (Interlink Network Systems, East Brunswick, New Jersey)
Xianlian (Alex) Zhou (CFD Corp, Huntsville, Alabama)
Xiaolin Man, PhD (Third Wave Systems, MN)
Yu Wei, PhD (Ford Motor Company) Harn-Jou Yeh, PhD (Taiwan)
Zan Mi, PhD (Caterpillar)

Past Summer Interns

Ben Goerdt (VSR)
Laith Qubain (High school)
Logan Butler (Iowa City West High School, Iowa City, IA)
Prajwal Kedilaya
Shacoya Smith (B.S. Program, Biomedical Engineering, The University of Iowa)
Riley Chapleau
Rob Lineback (West High School, Iowa City, IA)
Trevor Clinkenbeard
Zeid Qubain (High school)

Visiting Faculty and Scientists

Esteban Pena, PhD (Professor, Spain)
Hyung-Joo Kim, PhD (Researcher, Hyundai, Korea)
Joao Cardoso, PhD (Professor, Portugal)
Xuemei Feng, PhD (Professor, China)
Mohamad Alkam, PhD (Professor, JUST, Jordan)
Nasri Rabadi, PhD (Professor, AUM, Jordan)

Introduction

I find that the harder I work, the more luck I seem to have.
Thomas Jefferson (1743–1826)

1.1 What is predictive dynamics?

Predictive dynamics is a term coined to characterize a new methodology for predicting human motion while considering dynamics of the human and the environment...the laws of physics must be obeyed. Whereas *kinematics* is the study of motion (position, velocity, and acceleration) without forces and torques, *dynamics* is the study of motion of bodies with all external and internal forces taken into consideration.

For every motion affected by physics, there are laws that govern that motion. These laws have undergone the test of time, have sent people to the moon, and have been implemented into every computer program that calculates the motion of a given system. Equations that represent motion are called the equations of motion. For large redundant systems such as the human body, these equations become very complicated nonlinear coupled differential equations subject to various algebraic constraints on the motion; hence the term often used to represent these equations is differential algebraic equations (DAEs).

For a complicated system of segments, such as for a human being, which is made up of joints and rigid links (we assume rigidity of human bones), the formulation for multi-body dynamics becomes large and complex. Solving the consequent system of equations is almost impossible. Indeed, for structural systems with limited number of degrees of freedom, integrators have been developed to solve the problem iteratively. For a highly nonlinear system with a large number of degrees of freedom, however, integrators come to a halt.

For many years, our research at the Virtual Soldier Research (VSR) program has focused on a method that employs optimization with dynamics to predict human motion. Recent results have demonstrated significant promise for resolving the problem of predicting human motion while taking into consideration external forces, obstacles, physiology, and most importantly the equations of motion. Inducing natural behavior is a natural result of this method.

This method, which we call predictive dynamics (PD), provides a way to address the issue of predicting human motion in a general manner. It addresses problems where both the motion of the system and the forces causing the motion are unknown and must be determined simultaneously. It will be seen that an optimization-based formulation is ideally suited to solve such problems. Recent results have shown that this method is applicable to gait prediction, lifting movements, pushing and pulling movements, climbing, and many other tasks. Indeed, an entire task made up of multiple sub-tasks can be created whereby a true physics-based predictive human simulator is created.

The objective of this book is to clearly demonstrate the basic formulation needed to develop a PD task.

1.2 How does predictive dynamics work?

Predictive dynamics is an optimization-based method for predicting human motion, while taking into consideration the biomechanics, physics of the motion, and human behavior.

Consider a general optimization problem, for which there are three main ingredients (Arora, 2012):

1. A set of design variables, which in our case are the joint profiles (i.e., joint angles as functions of time) and the torque profiles at each joint
2. Multiple cost functions to be optimized, which are human performance measures that represent functions that are important to accomplishing the motion (e.g., energy, speed, joint torque, etc.), and
3. Constraints on the motion (e.g., collision avoidance, joint ranges of motion, etc.).

This general optimization problem is readily solved using existing optimization algorithms and codes. The field of optimization is mature and many such codes exist that have been verified and validated, and tested with many different complex problems.

Solving the above optimization problem predicts human motion as illustrated in Figure 1.1. It has been shown that for static postures (i.e., predicting final human postures to reach an object), this method is very successful; it produces human-like results (Abdel-Malek et al., 2001a–d; Abdel-Malek et al., 2004a–c; Abdel-Malek et al., 2006).

Now we add the issue of dynamics. We are interested in seeing how human motion is predicted for scenarios that involve dynamic influences including but not limited to external loads, obstacles, and running. The general concept is the addition of the laws of physics, i.e., equations of motion, as constraints. Instead of calculating specific static postures, we now calculate time-varying angles for each joint in the body, which are also called motion profiles. Instead of a simple displacement cost function, we implement an energy and effort driving performance measures, which drive the motion to minimize these two measures. This is indeed the essence of our theoretical framework...we believe that humans act and move because humans want to minimize or maximize certain objectives. This

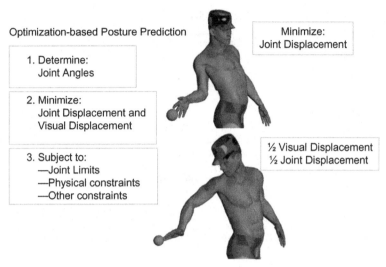

FIGURE 1.1

Santos® the human model reaches to a point using two different strategies as a result of predicting postures.

approach yields natural human motion, induces behavior, and provides for a straightforward method of "predicting" the dynamics of the motion.

Figure 1.2 depicts the general optimization-based algorithm. The goal is to determine joint profiles, meaning the angles subtended by each body segment as a function of time. The second optimization formulation ingredient is one or more cost functions. In this case, we have selected energy and effort, assuming that both of these human performance measures can be transcribed into mathematical functions. The third ingredient is the same as shown in Figure 1.1; however, we have added an additional important constraint, namely the equations of motion. These are the laws of motion that govern how our world behaves dynamically.

In general, we consider any case where a human segment is undergoing motion that warrants the consideration of masses and moments of inertia. PD can incorporate such general cases.

1.3 Why data-driven human motion prediction does not work

We firmly believe that the data-driven approach to human motion prediction is the wrong approach. Thousands of experiments are typically done to capture a few motions. These motions are then compiled into large tables with many para-meters. The data is then analyzed and modeled as a nonlinear or functional

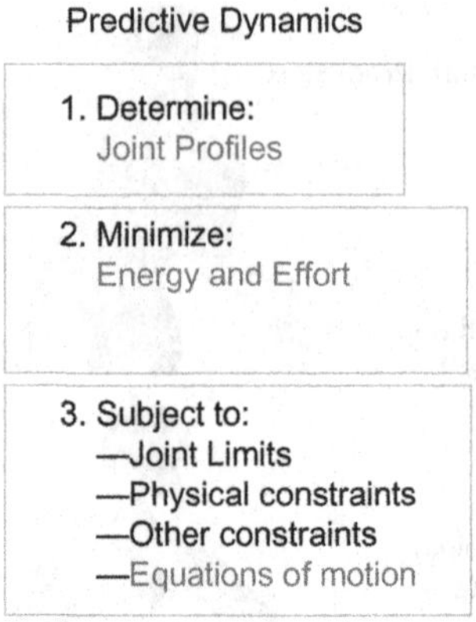

Predictive Dynamics

1. Determine:
   Joint Profiles

2. Minimize:
   Energy and Effort

3. Subject to:
   —Joint Limits
   —Physical constraints
   —Other constraints
   —Equations of motion

FIGURE 1.2

The general formulation for PD illustrated as a three-step optimization formulation.

regression model that should, in principle, predict motion. There are many obvious problems with this method, including:

- Difficulties in collecting the data for varying anthropometries. This includes the changes of masses, moments of inertia, muscle performance, and many other parameters for each person.
- Difficulties in managing a large number of parameters in a functional or nonlinear regression algorithm. A large number of parameters means a complex and less accurate model.
- Difficulties in predicting postures and motions for reaches that have obstacles. For each obstacle, the experiments must be repeated.
- Difficulties in predicting motions where dynamics (external forces and loads) play an important role.

After the apple fell from the tree on Newton's head, he proceeded to measure a few more, came up with the general theory, and finally came up with a rigorous mathematical formulation for all falling objects and, furthermore, for all objects in motion. He did not measure every apple on every tree to come up with a theory that works and that is the fundamental basis for all motion in our universe.

The idea of recording every motion for thousands of people and for thousands of different scenarios does provide a good way to study motion and to validate motion predicted with various methods. However, it has no value for the prediction of motions beyond static postures.

1.4 Concluding remarks

This book deals with the science of human motion. It presents a rigorous methodology for representing human biomechanics and joint motion, and includes the effects of environmental physics on the motion. PD is a novel approach for simulating, specifically *predicting*, human motion. It avoids direct integration of differential-algebraic equations in order to create the resulting simulations for redundant digital human models. Cause and effect is at the center of this formulation...using a digital human environment, a user is able to model the human by selecting their anthropometry, body type, weight, strength, and fatigue limits. The user is also able to load the digital human model with various loads, for example perhaps inflict a biomechanical injury that would restrict a joint range of motion or lower the strength value of a particular group of muscles. Upon selecting a motion or a task, PD provides for a computational platform that lets us know the human reaction to these conditions...indeed, it answers the question: "How would the human have reacted if they were under these conditions?" It is a human simulator.

The book aims to illustrate the entire methodology beginning with a systematic method for modeling the kinematics, then creating an optimization formulation, writing the dynamics, and formulating the PD problem. Human performance measures are also introduced as cost functions (objective functions) that drive the motion, a theory that has been proven effective for producing natural motions. Such motion prediction capabilities have a wide variety of practical applications, such as in automotive industry, military, clinical and biomechanical analyses, and design of equipment. Several such applications are illustrated through detailed examples.

References

Abdel-Malek, K., Yu, W., Jaber, M., 2001a. Realistic Posture Prediction. 2001 SAE Digital Human Modeling and Simulation.

Abdel-Malek, K., Wei, Y., Mi, Z., Tanbour, E., Jaber, M., 2001b. Posture prediction versus inverse kinematics. In: Proceedings of the 2001 ASME Design Engineering Technical Conferences and Computers and Information in Engineering Conference, Pittsburgh, PA, pp. 37−45.

Abdel-Malek, K., Yang, J., Brand, R., Tanbour, E., 2001c. Towards understanding the workspace of the upper extremities. SAE Trans. J. Passenger Cars: Mech. Syst. 110 (6), 2198−2206.

Abdel-Malek, K., Yu, W., Jaber, M., Duncan, J., 2001d. Realistic posture prediction for maximum dexterity. SAE Technical Paper 2001-01-2110. doi: 10.427/2001-01-2110.

Abdel-Malek, K., Yang, J., Brand, R., Tanbour, E., 2004a. Towards understanding the workspace of human limbs. Ergonomics 47 (13), 1386−1406.

Abdel-Malek, K., Yang, J., Yu, W., Duncan, J., 2004b. Human performance measures: mathematics. Proceedings of the ASME Design Engineering Technical Conferences (DAC 2004), Salt Lake City, UT.

Abdel-Malek, K., Yang, J., Mi, Z., Patel, V.C., Nebel, K., 2004c. Human upper body motion prediction. Proceedings of Conference on Applied Simulation and Modeling (ASM). Rhodes, Greece, pp. 28−30.

Abdel-Malek, K., Yang, J., Marler, T., Beck, S., Mathai, A., Zhou, X., Patrick, A., Arora, J.S., 2006. Towards a new generation of virtual humans. International Journal of Human Factors Modelling and Simulation 1 (1), 2−39.

Arora, J.S., 2012. Introduction to Optimum Design, third ed. Elsevier, Inc., Waltham, MA, USA.

Human Modeling: Kinematics

The optimist proclaims that we live in the best of all possible worlds, and the pessimist fears this is true.
James Branch Cabell

2.1 Introduction

The objective of this chapter is to establish a systematic method for representing human anatomy and to develop mathematical methods for kinematic analysis as the human body undergoes motion. Kinematic analysis in this context means the study of motion characterized by the position, velocity, and acceleration of human segmental links.

Throughout this chapter, we shall consider various segments of the body as individual rigid bodies that are connected via joints. Human modeling techniques have rapidly evolved in recent years, driven by the need for safety, security, and better ergonomics, as well as the need for avatars to perform tasks that could not be performed in the real world. Perhaps the most influential force behind this fast pace is the gaming industry where avatars are extensively used to interact and respond in real time. Similarly, in the movie industry, digital characters are used to replace actors where it has now become difficult in some cases to differentiate the real from the virtual. In general, human models have been represented as stick figures, skeletons, mesh surfaces, profiles, and mannequins (Figure 2.1).

Consider the motion of a person's arm from one position to another, where only the elbow joint is changed. As a result of this simple motion, the hand is also moved in space to a final configuration defined by a position and an orientation (also called a pose). In order to characterize the motion of these segmental links and their associated joints, it is necessary to establish a systematic approach for specifying coordinate systems defined by *xyz* on each link, and establish a method for relating any two such coordinate systems.

Let us assume for a moment that we are able to specify values of the joint variables for the shoulder complex, the elbow, and the wrist, and we wanted to know the final position of the hand (Figure 2.2). In this case, we specify a vector **x** that describes the position of the hand with respect to another coordinate system

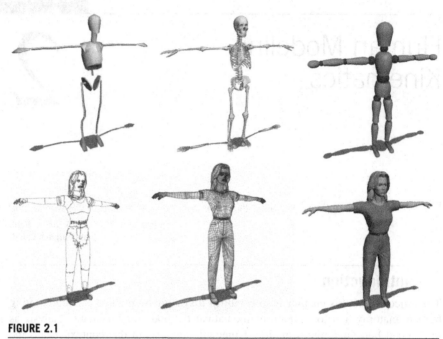

FIGURE 2.1

Digital representations of human models.

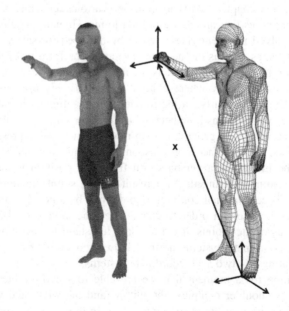

FIGURE 2.2

A vector **x** representing the position of the hand with respect to the foot.

(e.g., the foot). This chapter addresses exactly this issue. It will provide a rigorous method for formulating a set of equations that have the joint variables as their parameters. If the final hand position is required, variations in the joint variables are substituted into the equation and the final position is readily obtained.

We shall also introduce a method for modeling human joints, as simple or as complex as necessary, that represents the resulting interaction between any two segmental links. The simplest form of these joints is the rotational joint (such as the elbow joint). The combination of a number of simple joints can become complex in nature but still be represented using this straightforward approach.

The Denavit−Hartenberg (DH) method was created in the 1950s to systematically represent the relation between two coordinate systems, but was only extensively used in the early 1980s with the appearance of computational methods and hardware that enable the necessary calculations. The method is currently used to a great extent in the analysis and control of robotic manipulators. This method has also been successful in addressing human motion, in particular towards gaining a better understanding of the mechanics of human motion.

It is important to distinguish the difference between a rigid body and a flexible body. A rigid body is one that cannot deform (we typically consider bone as non-deforming, at least for the moment). However, a flexible body (or deformable object) is one that undergoes relatively large strains when subjected to a load (e.g., soft tissue). For the approach presented in this section, only rigid body motion is assumed at all times. Indeed, for ergonomic design considerations, rigid body motion is adequate to address most problems. Muscle interaction and deformation will be addressed in later chapters.

The DH modeling method is suitable for addressing the motion of kinematic structures that are arranged in series. The DH method will be used to perform analysis on the human body in this chapter, and will be used to predict postures and perform ergonomic analysis in later chapters. *A posture is defined in this text as the configuration of a series of segmental links in the human body.*

The human body is indeed arranged in series, where each independent anatomical structure is connected to another via a joint. Consider, for example, that there exists a main coordinate system located at the waist. From that coordinate system, one may be able to draw a branch by identifying a rigid link, connected through a joint to another rigid link, connected to another link, until you reach the hand. Each finger also comprises a number of segmental links connected via joints. Similarly, also starting from the waist, one may follow the connection to reach the head, the other hand, the left foot, and the right foot. We shall refer to one such chain as a branch. For example, Figure 2.3 depicts the modeling of a human into a number of kinematic branches. A hand can be represented by five branches, one for each finger.

Because we consider gross human motion, detailed modeling of the joint connection is less important at this stage. Nevertheless, because in many cases accurate biomechanical modeling of a joint is needed, in Chapter 4 we will also present a more elaborate method for representing the kinematic interactivity

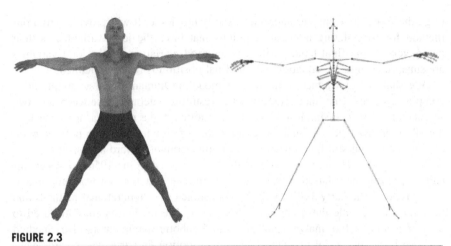

FIGURE 2.3

Modeling of a human using a series of rigid links connected by joints, also called kinematic skeleton.

within the joint while taking into consideration muscle action, ligaments, and other effects that are not considered in the DH representation method.

This chapter will begin with the fundamental theories required to understand motion in 3D. The general translational and rotational motion of an object will first be presented, followed by a standardization of a method for embedding coordinate systems (also called triads) in each segmental link. We will then develop a formulation that relates any two segmental links in this chain.

2.2 General rigid body displacement

We define the word *configuration* as denoting the position and orientation of a rigid body. Consider the general motion of the hand, now considered as a rigid body, where the line segment $\overline{OW}$ embedded in the hand is shown in Figure 2.4.

The motion will carry the rigid body from its initial configuration in coordinate system A to a different configuration indicated by the line segment $\overline{O''W''}$. This motion can be described in vector notation as a translation along the vector $\mathbf{p} = \overline{OO'}$ and a rotation about O' prescribed by the rotation matrix $^A\mathbf{R}_B$, where this rotation matrix rotates the A-coordinate system from its orientation to the orientation shown in the B-coordinate system. In vector notation, this motion can be described as

$$^A\mathbf{x}(W'') \equiv \overline{OO'} + \overline{O'W''} \tag{2.1}$$

where $^A\mathbf{x}(W'')$ denotes the vector extending from the origin O to W'' as seen by the A-coordinate system. A superscript to the left of the letter denotes

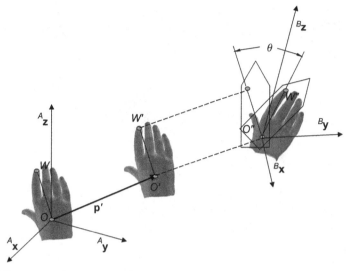

FIGURE 2.4

General motion of a rigid body from one position and orientation at *A* to a second position and orientation at *B*.

the reference frame in which the vector is resolved and is read as: **x** as seen by *A*. This general motion can also be written in terms of a rotation matrix as

$$^A\mathbf{x} = {^A}\mathbf{p} + {^A}\mathbf{R}_B{^B}\mathbf{x} \tag{2.2}$$

where $^B\mathbf{x}$ is the vector $\overline{O''W''}$ resolved in the *B*-coordinate system.

Of course, it is also possible that the rigid body undergoes first a rotation followed by a translation, the results of which are the same. Equation (2.2) is indeed the most important result, which applies to the general motion of a rigid body in 3D space.

2.2.1 Example: rotation and translation

Calculating the effects of translations and rotations on target points and vectors in a virtual environment is an important aspect. By inserting a virtual camera into the human's eyes, the digital human is able to report back what can be seen from this location, whether there exists obstacles in the design of a vehicle, but more importantly, we have to calculate how the target point is seen by the original coordinate system before motion. Consider a target point Q shown in Figure 2.5, and given by $^A\mathbf{x}_Q = [1 \quad 0]^T$, which is read as a vector of magnitude 1 along the x-axis and 0 along the y-axis, as resolved (or seen) by the *A*-coordinate system. The person now walks to a new position (at 5 along the x-axis and 2 along the y-axis), which therefore can be represented by a vector from the origin of the

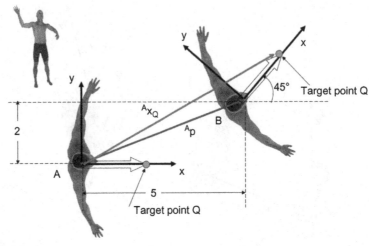

FIGURE 2.5

Calculating the coordinates of the target point Q after rotation and translation from one configuration to another.

A-coordinate system as $^A\mathbf{p} = [5 \quad 2]^T$. The person also rotates an angle of 45° (line of sight by 45° CCW). It is required to calculate the vector describing the final target point with respect to the A-coordinate system.

The rotation matrix representing this rotation can be written as the dot product of the B-coordinate system with the A-coordinate system.

$$^A\mathbf{R}_B = \begin{bmatrix} ^A\mathbf{x} \cdot ^B\mathbf{x} & ^A\mathbf{x} \cdot ^B\mathbf{y} \\ ^A\mathbf{y} \cdot ^B\mathbf{x} & ^A\mathbf{y} \cdot ^B\mathbf{y} \end{bmatrix}$$

$$^A\mathbf{R}_B = \begin{bmatrix} 0.707 & -0.707 \\ 0.707 & 0.707 \end{bmatrix} \tag{2.3}$$

The vector describing the coordinates of the target point Q after rotation and translation, and as seen by the original coordinate system A, is calculated as

$$^A\mathbf{x}_Q \equiv {}^A\mathbf{p} + {}^A\mathbf{R}_B {}^B\mathbf{x}_Q = \begin{bmatrix} 5 \\ 2 \end{bmatrix} + {}^A\mathbf{R}_B = \begin{bmatrix} 0.707 & -0.707 \\ 0.707 & 0.707 \end{bmatrix} \begin{bmatrix} 1 \\ 0 \end{bmatrix} = \begin{bmatrix} 5.707 \\ 2.707 \end{bmatrix} \tag{2.4}$$

It is now evident that the rotation matrix plays an important role in various ways.

a. The rotation matrix can be used to describe the orientation of a set of vectors in one coordinate system to another.

b. The rotation matrix can be used to calculate the coordinates of a point after a rotation and translation of coordinate systems.

2.3 Concept of extended vectors and homogeneous coordinates

The concept of an extended vector is introduced to facilitate vector and matrix operations. The extended vector of $\tilde{x} = [x \quad y \quad z]^T$ is the (4×1) vector given by $x = [ax \quad ay \quad az \quad a]^T$ where a is an arbitrary number. The concept of *homogeneous coordinates* is introduced indicating that the values become homogeneous when adding a new coordinate to the vector. In this case, it is homogeneous because the vector will have the same meaning even if multiplied by a constant. Formulations involving homogeneous coordinates are often simpler and more symmetric than their Cartesian counterparts. Homogeneous coordinates have many applications, including computer graphics and 3D computer vision, where affine transformations are allowed and projective transformations are easily represented by a matrix. We shall use the concept of a homogeneous transformation to represent the rotation and translation into one homogeneous matrix transformation.

In human modeling, the importance of using homogeneous coordinates and the concept of an extended vector stem from the representation of Equation (2.2), which will become fundamental to the formulation of a systematic method for representing the motion of one segmental link with respect to another. It is possible to write Equation (2.2) in terms of a (4×4) matrix as

$$\begin{bmatrix} ^A\tilde{x} \\ \hline 1 \end{bmatrix} \equiv \begin{bmatrix} ^A\mathbf{R}_B & \vdots & ^A\mathbf{p} \\ \hline \mathbf{0} & \vdots & 1 \end{bmatrix} \begin{bmatrix} ^B\tilde{x} \\ \hline 1 \end{bmatrix} \tag{2.5}$$

where the extended vectors $^A x = \begin{bmatrix} ^A\tilde{x} \\ \hline 1 \end{bmatrix}$ and $^B x = \begin{bmatrix} ^B\tilde{x} \\ \hline 1 \end{bmatrix}$ can be used to rewrite Equation (2.3) as

$$^A x = \begin{bmatrix} ^A\mathbf{R}_B & \vdots & ^A\mathbf{p} \\ \hline \mathbf{0} & \vdots & 1 \end{bmatrix} {}^B x \tag{2.6}$$

or as

$$^A x = {}^A\mathbf{T}_B {}^B x \tag{2.7}$$

where

$$^A\mathbf{T}_B = \begin{bmatrix} ^A\mathbf{R}_B & \vdots & ^A\mathbf{p} \\ \hline \mathbf{0} & \vdots & 1 \end{bmatrix} \tag{2.8}$$

This $^A\mathbf{T}_B$ matrix is called the Homogenous Transformation matrix and is read as the transformation from B to A. It can be seen as acting on a rigid body causing a transformation, i.e., changing its configuration. On the other hand, and this is the concept that will be used throughout this text, it is seen as an operator acting on a vector $^B x$ (which is resolved in the B-coordinate system), and resolving the resulting vector in the A-coordinate system. This is similar in action to the rotation matrix but includes the translation as well.

The transformation matrix is indeed a partitioned matrix where the upper left corner sub-matrix is the rotation matrix and the upper right vector is the position vector

$$
{}^A\mathbf{T}_B = \left[\begin{array}{c|c} \text{rotation} & \text{position} \\ \text{matrix} & \text{vector} \\ \hline \mathbf{0} & 1 \end{array}\right] \tag{2.9}
$$

2.4 Basic transformations

Figure 2.6 shows two coordinate systems $x_1y_1z_1$ and $x_2y_2z_2$ that are coincident. Consider the matrix generated by the rotation of the coordinates system $x_2y_2z_2$ about the z_1 axis.

By definition, this rotation can be written as the dot product of unit vectors as

$$
\mathbf{R}_{z,\theta} = \begin{bmatrix} x_1 \cdot x_2 & x_1 \cdot y_2 & x_1 \cdot z_2 \\ y_1 \cdot x_2 & y_1 \cdot y_2 & y_1 \cdot z_2 \\ z_1 \cdot x_2 & z_1 \cdot y_2 & z_1 \cdot z_2 \end{bmatrix} \tag{2.10}
$$

where the subscripts z and θ denote a rotation about z with an angle θ. From Figure 2.6(B), carrying out the dot product yields

$$
\mathbf{R}_{z,\theta} = \begin{bmatrix} \cos\theta & \cos(90 + \theta) & 0 \\ \cos(90 - \theta) & \cos\theta & 0 \\ 0 & 0 & 1 \end{bmatrix} \tag{2.11}
$$

Further simplification using the identity $\cos(90 + \theta) = -\sin\theta$ and $\cos(90 - \theta) = \sin\theta$ yields the basic rotation matrix

$$
\mathbf{R}_{z,\theta} = \begin{bmatrix} \cos\theta & -\sin\theta & 0 \\ \sin\theta & \cos\theta & 0 \\ 0 & 0 & 1 \end{bmatrix} \tag{2.12}
$$

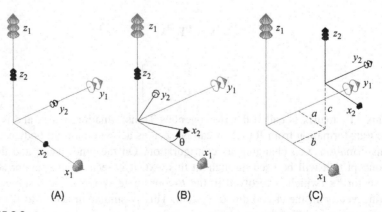

(A) (B) (C)

FIGURE 2.6

Definition of basic transformations.

Inserting this basic rotation matrix into the transformation matrix **T** with no displacement yields (i.e., the first three elements of the last column are all zeros)

$$\mathbf{T}_{z,\theta} = \begin{bmatrix} \cos\theta & -\sin\theta & 0 & 0 \\ \sin\theta & \cos\theta & 0 & 0 \\ 0 & 0 & 1 & 0 \\ 0 & 0 & 0 & 1 \end{bmatrix} \qquad (2.13)$$

which is a *basic homogeneous transformation* matrix for rotation about the z-axis.

Similarly, two other basic homogeneous transformation matrices for rotation about the x- and y-axes, respectively, are given by

$$\mathbf{T}_{x,\alpha} = \begin{bmatrix} 1 & 0 & 0 & 0 \\ 0 & \cos\alpha & -\sin\alpha & 0 \\ 0 & \sin\alpha & \cos\alpha & 0 \\ 0 & 0 & 0 & 1 \end{bmatrix} \qquad (2.14)$$

and

$$\mathbf{T}_{y,\varphi} = \begin{bmatrix} \cos\varphi & 0 & \sin\varphi & 0 \\ 0 & 1 & 0 & 0 \\ -\sin\varphi & 0 & \cos\varphi & 0 \\ 0 & 0 & 0 & 1 \end{bmatrix} \qquad (2.15)$$

A pure translation matrix, called a basic transformation matrix for translation, can be written not by specifying rotations (i.e., an identity matrix for the rotation matrix), but by specifying the coordinates of the three elements in the last column. A transformation matrix for translation along the x-axis by a-units can be written as

$$\mathbf{T}_{x,a} = \begin{bmatrix} 1 & 0 & 0 & a \\ 0 & 1 & 0 & 0 \\ 0 & 0 & 1 & 0 \\ 0 & 0 & 0 & 1 \end{bmatrix} \qquad (2.16)$$

Similarly, a basic transformation matrix translating along x-, y-, and z-axes is written as

$$\mathbf{T}_{translation} = \begin{bmatrix} 1 & 0 & 0 & a \\ 0 & 1 & 0 & b \\ 0 & 0 & 1 & c \\ 0 & 0 & 0 & 1 \end{bmatrix} \qquad (2.17)$$

which can be thought of as translating the $x_2 y_2 z_2$ a distance a along the x-axis, a distance b along the y-axis, and a distance c along the z-axis as shown in Figure 2.6(C).

2.4.1 **Example: knee rotation**

The knee joint is a pivotal hinge joint, which permits flexion and extension as well as a slight medial and lateral *rotation*. For the purpose of this example we shall consider it as purely a hinge joint with one degree of freedom (DOF). We seek to represent the position vector of a point specified on the foot as the lower limb undergoes a rotational motion at the knee. A point Q on the foot is shown in Figure 2.7 and is represented by the vector $\mathbf{x}_Q = [3 \quad -7 \quad 0 \quad 1]^T$ with respect to the coordinate system located at the knee. The knee is constrained to allow motion only about the axis $\mathbf{z}$ by an angle θ. It is required to calculate the final position of the foot point Q as the joint rotates (i.e., as a function of θ).

The lower limb rotates about the axis $\mathbf{z}$, thus the pure rotational homogenous matrix is

$$\mathbf{T}_{z,\theta} = \begin{bmatrix} \cos\theta & -\sin\theta & 0 & 0 \\ \sin\theta & \cos\theta & 0 & 0 \\ 0 & 0 & 1 & 0 \\ 0 & 0 & 0 & 1 \end{bmatrix} \tag{2.18}$$

Note the translation vector is zero. To determine the location of Q at any joint displacement θ, we multiply $\mathbf{x}_Q$ by $\mathbf{T}_{z,\theta}$ to calculate the rotated vector $\mathbf{x}'_Q$

$$\mathbf{x}'_Q = \mathbf{T}_{z,\theta}\mathbf{x}_Q = [3\cos\theta+7\sin\theta \quad 3\sin\theta-7\cos\theta \quad 0 \quad 1]^T \tag{2.19}$$

With this expression, it is possible to calculate the position of the lower limb at any specified value of θ. For example let $\theta = 90°$, then the rotated limb's new position is

$$\mathbf{x}'_Q(90°) = \begin{bmatrix} 7 & 3 & 0 & 1 \end{bmatrix}^T \tag{2.20}$$

which is shown in Figure 2.7 as the lower limb is extended.

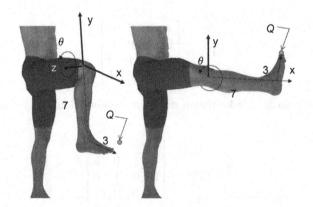

FIGURE 2.7

Rotation of the lower limb about the z-axis by an angle θ.

2.5 **Composite transformations**

In this section, we show that two consecutive transformation matrices can be characterized by the multiplication of two independent transformation matrices that yield the same result. In fact, we will show that consecutive transformations can be represented by the multiplication of their respective transformation matrices. Indeed, consider the transformation from B to A given by the matrix $^A\mathbf{T}_B$. Consider also another transformation from B to C represented by the transformation matrix $^B\mathbf{T}_C$. The resulting transformation can be obtained by multiplying the two transformation matrices as

$$^A\mathbf{T}_B\,^B\mathbf{T}_C = \left[\begin{array}{c|c} ^A\mathbf{R}_B & ^A\mathbf{p}_{AB} \\ \hline 000 & 1 \end{array}\right]\left[\begin{array}{c|c} ^B\mathbf{R}_C & ^B\mathbf{p}_{BC} \\ \hline 000 & 1 \end{array}\right] \tag{2.21}$$

$$^A\mathbf{T}_B\,^B\mathbf{T}_C = \left[\begin{array}{c|c} ^A\mathbf{R}_B\,^B\mathbf{R}_C + 0 & ^A\mathbf{R}_B\,^B\mathbf{p}_{BC} + ^A\mathbf{p}_{AB} \\ \hline 000 & 1 \end{array}\right] \tag{2.22}$$

The multiplication of two consecutive rotation matrices $^A\mathbf{R}_B\,^B\mathbf{R}_C$ yields the rotation matrix $^A\mathbf{R}_C$. In addition, the position vector to the origin of the B coordinate system is seen by the A coordinate system as

$$^A\mathbf{R}_B\,^B\mathbf{p}_{BC} = ^A\mathbf{p}_{BC} \tag{2.23}$$

and the summation of the two position vectors yields

$$^A\mathbf{p}_{BC} + ^A\mathbf{p}_{AB} = ^A\mathbf{p}_{AC} \tag{2.24}$$

Substituting Equations (2.23 and 2.24) into Equation (2.22) yields

$$^A\mathbf{T}_B\,^B\mathbf{T}_C = \left[\begin{array}{c|c} ^A\mathbf{R}_C & ^A\mathbf{p}_{AC} \\ \hline 000 & 1 \end{array}\right] = ^A\mathbf{T}_C \tag{2.25}$$

which indicates that the chain rule applied to rotation matrices is also applicable to transformation matrices. In analogy with rotation matrices, the concept of extending the chain rule to a sequence of transformations can be applied such that the resulting transformation matrix characterizes the combined motion from A to C. This is a very important result, which will be used extensively throughout this text.

2.5.1 **Example: composite transformations**

Consider the arm shown in Figure 2.8 with three coordinate systems. For this model, there exists one coordinate system at the shoulder called A, one at the elbow called B, and one at the wrist called C. Note that all coordinate systems are restricted and can only rotate about their own z-axis.

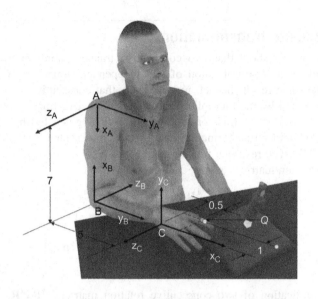

FIGURE 2.8

The upper limb with three embedded coordinate systems at the shoulder, elbow, and hand.

A point Q on the thumb is represented by a vector embedded in the hand and given by $^{C}\mathbf{x}_{Q} = \begin{bmatrix} 0.5 & 0 & -1 & 1 \end{bmatrix}^{T}$. It is required to

(a) Determine the homogenous transformation matrix $^{A}\mathbf{T}_{B}$.
(b) Determine the homogenous transformation matrix $^{B}\mathbf{T}_{C}$.
(c) Determine the homogenous transformation matrix $^{A}\mathbf{T}_{C}$.
(d) Determine the coordinates of the vector $\mathbf{x}_{Q}$ with respect to the shoulder coordinate system A (i.e., $^{A}\mathbf{x}_{Q}$).

The transformation matrix relating coordinate systems A to B is

$$^{A}\mathbf{T}_{B} = \begin{bmatrix} -1 & 0 & 0 & 7 \\ 0 & 1 & 0 & 0 \\ 0 & 0 & -1 & 0 \\ 0 & 0 & 0 & 1 \end{bmatrix} \tag{2.26}$$

$$^{B}\mathbf{T}_{C} = \begin{bmatrix} 0 & 1 & 0 & 0 \\ 1 & 0 & 0 & 5 \\ 0 & 0 & -1 & 0 \\ 0 & 0 & 0 & 1 \end{bmatrix} \tag{2.27}$$

In order to determine the position and orientation of the C coordinate system with respect to the A coordinate system, we multiply the two matrices

$$^A\mathbf{T}_B\,^B\mathbf{T}_C = \begin{bmatrix} 0 & -1 & 0 & 7 \\ 1 & 0 & 0 & 5 \\ 0 & 0 & 1 & 0 \\ 0 & 0 & 0 & 1 \end{bmatrix} \tag{2.28}$$

To determine how the thumb is seen by coordinate system A, we calculate

$$^A\mathbf{x}_Q = {}^A\mathbf{T}_B\,^B\mathbf{T}_C\,^C\mathbf{x}_Q = \begin{bmatrix} 7 & 5.5 & -1 & 1 \end{bmatrix}^T \tag{2.29}$$

From Figure 2.8, it can be observed that indeed the vector describing the point on the thumb has coordinates $(7, 5.5, -1)$ with respect to the A-coordinate system.

2.6 Directed transformation graphs

Consider the coordinate frames depicted in Figure 2.9. Each graph from one frame to another represents a transformation matrix and is denoted by the **T**-matrix. The direction of the arrow indicates subscript and superscript, respectively, of the **T**-matrix, i.e., the transformation from frame 0 to frame 1 is denoted by $^0\mathbf{T}_1$. A graph from frame 1 to frame 2 is represented by $^1\mathbf{T}_2$.

Applying a sequence of transformations such as $^0\mathbf{T}_1$ followed by $^1\mathbf{T}_2$ yields a graph from frame 0 directly to frame 2 represented by the transformation matrix

$$^0\mathbf{T}_2 = {}^0\mathbf{T}_1\,^1\mathbf{T}_2 \tag{2.30}$$

Similarly, applying another transformation from frame 2 to frame 3 can be represented by a directed graph from frame 0 to frame 3 and characterized by

$$^0\mathbf{T}_3 = {}^0\mathbf{T}_1\,^1\mathbf{T}_2\,^2\mathbf{T}_3 \tag{2.31}$$

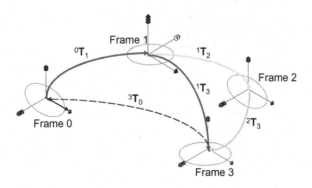

FIGURE 2.9

Four directed transformation graphs.

A graph from frame 3 to frame 0 is represented by $^3\mathbf{T}_0$; therefore, the multiplication of a transformation matrix times its inverse yields the identity matrix

$$^0\mathbf{T}_3\,^3\mathbf{T}_0 = \mathbf{I} \tag{2.32}$$

It is also noted here that a transformation between any two frames can be obtained independent of the route followed, i.e., the transformation from frame 0 to frame 2 can be obtained by

$$^0\mathbf{T}_2 = {}^0\mathbf{T}_1\,^1\mathbf{T}_2 \tag{2.33}$$

or by

$$^0\mathbf{T}_2 = {}^0\mathbf{T}_3\,^3\mathbf{T}_2 \tag{2.34}$$

where the transformation

$$^3\mathbf{T}_2 = (^2\mathbf{T}_3)^{-1} \tag{2.35}$$

The order in which matrices are multiplied is important since matrix multiplication is not commutative, i.e., in general $^A\mathbf{T}_B\,^B\mathbf{T}_C \neq {}^B\mathbf{T}_C\,^A\mathbf{T}_B$. In fact, great care must be given to the order of multiplication. Consider two coordinate systems, $X_1Y_1Z_1$ being the world coordinate system, and $X_2Y_2Z_2$ being the body reference frame. Two rules must be followed in applying the order of multiplication of transformation matrices. These rules are given without proof:

1. A transformation taking place with respect to the world reference frame ($X_1Y_1Z_1$) necessitates the *pre-multiplication* of the previous transformation matrix by an appropriate basic homogeneous transformation matrix.
2. A transformation taking place with respect to the body's own reference frame ($X_2Y_2Z_2$) necessitates the post-multiplication of the previous transformation matrix by an appropriate basic transformation matrix.

2.6.1 Example: multiple transformations

A digital human lives in a computer-aided engineering environment. This human will be requested to perform some tasks, i.e., to grasp and move objects. In order to identify objects in the workspace, a virtual camera is embedded in the human's head, which will function as his eyes. This camera will determine the position and orientation of an object in space and will return a homogeneous transformation matrix. The virtual camera senses the position of a Joystick $\mathbf{J}$ shown in Figure 2.10. The virtual camera's coordinate system is represented by $\mathbf{c}_1$, $\mathbf{c}_2$, and $\mathbf{c}_3$ as shown in Figure 2.10. The camera identifies the position and configuration of the shoulder and the joystick.

The homogeneous transformation matrix of the joystick $\mathbf{J}$ as seen by the camera $\mathbf{C}$ is represented by $^C\mathbf{T}_J$ as

$$^C\mathbf{T}_J = \begin{bmatrix} 0 & 1 & 0 & 5 \\ 0 & 0 & 1 & 24 \\ 1 & 0 & 0 & -15 \\ 0 & 0 & 0 & 1 \end{bmatrix} \tag{2.36}$$

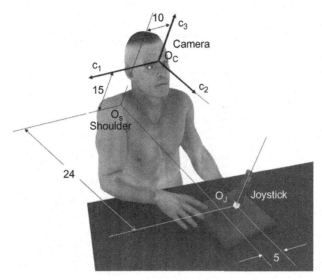

FIGURE 2.10

A digital human whose eyes are replaced by a camera.

The transformation matrix of the coordinate system embedded in the shoulder with respect to the camera is given as

$$^C\mathbf{T}_S = \begin{bmatrix} 0 & 0 & 1 & 10 \\ 1 & 0 & 0 & 0 \\ 0 & 1 & 0 & -15 \\ 0 & 0 & 0 & 1 \end{bmatrix} \tag{2.37}$$

It is required to:

i. Sketch the unit basis vectors $\mathbf{j}_1$, $\mathbf{j}_2$, and $\mathbf{j}_3$ in the correct orientation at the origin O_j of the joystick **J**.
ii. Sketch the unit basis vectors $(\mathbf{s}_1, \mathbf{s}_2, \mathbf{s}_3)$ in the correct orientation at the origin O_S of the shoulder **S**.
iii. Calculate the coordinates of the Joystick origin O_j relative to the shoulder **S**.
iv. Determine the transformation matrix $^S\mathbf{T}_H$ for the hand **H** as seen by the shoulder, when the hand is positioned to grasp the cube, as shown in Figure 2.11.

With regards to (i) and (ii) above, since the rotation matrix $^C\mathbf{R}_J$ extracted from $^C\mathbf{T}_J$ is given by

$$^C\mathbf{R}_J = \begin{bmatrix} 0 & 1 & 0 \\ 0 & 0 & 1 \\ 1 & 0 & 0 \end{bmatrix} \tag{2.38}$$

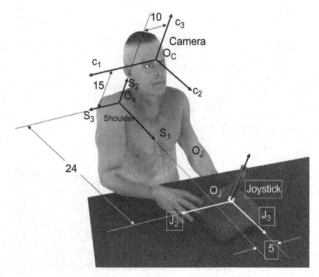

FIGURE 2.11

Orienting the hand to grasp an object.

then vector bases at O_C are given by

$$[\mathbf{j}_1 \mid \mathbf{j}_2 \mid \mathbf{j}_3] = \begin{bmatrix} 0 & 1 & 0 \\ 0 & 0 & 1 \\ 1 & 0 & 0 \end{bmatrix} \tag{2.39}$$

which can readily be drawn at O_J with respect to the video camera coordinate system and are shown in Figure 2.11. Similarly, the unit basis $[\mathbf{s}_1 \mid \mathbf{s}_2 \mid \mathbf{s}_3]$ are extracted from $^C\mathbf{T}_S$ and are plotted as shown in Figure 2.11.

With regards to (iii) above, the coordinates of the joystick origin relative to the shoulder coordinate frame $\mathbf{S}$ (i.e., seeking the vector $^S\mathbf{p}_{SJ}$) can be either read from Figure 2.12 as $^S\mathbf{p}_{SJ} = \begin{bmatrix} 24 & 0 & -5 \end{bmatrix}^T$ or calculated numerically from $^S\mathbf{T}_J$, which can be written as

$$^S\mathbf{T}_J = {}^S\mathbf{T}_C \, {}^C\mathbf{T}_J \tag{2.40}$$

Since only $^S\mathbf{T}_C$ is given, its inverse is computed as

$$^S\mathbf{T}_C = [{}^C\mathbf{T}_S]^{-1} = \begin{bmatrix} 0 & 1 & 0 & 0 \\ 0 & 0 & 1 & 15 \\ 1 & 0 & 0 & -10 \\ 0 & 0 & 0 & 1 \end{bmatrix} \tag{2.41}$$

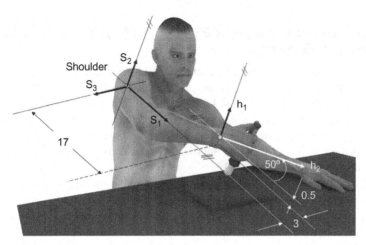

FIGURE 2.12

Solution to (i) and (ii) in Example 2.6.1.

Calculating $^S\mathbf{T}_J$ yields

$$
^S\mathbf{T}_J = \begin{bmatrix} 0 & 1 & 0 & 0 \\ 0 & 0 & 1 & 15 \\ 1 & 0 & 0 & -10 \\ 0 & 0 & 0 & 1 \end{bmatrix} \begin{bmatrix} 0 & 1 & 0 & 5 \\ 0 & 0 & 1 & 24 \\ 1 & 0 & 0 & -15 \\ 0 & 0 & 0 & 1 \end{bmatrix} = \begin{bmatrix} 0 & 0 & 1 & 24 \\ 1 & 0 & 0 & 0 \\ 0 & 1 & 0 & -5 \\ 0 & 0 & 0 & 1 \end{bmatrix} \tag{2.42}
$$

and extracting $^S\mathbf{p}_{SJ} = \begin{bmatrix} 24 & 0 & -5 \end{bmatrix}^T$ yields identical results.

With regards to (iv) above, the transformation matrix $^B\mathbf{T}_F$ can be first obtained by identifying the unit basis vectors with respect to the shoulder coordinate frame as

$$
^S\mathbf{R}_H = \begin{bmatrix} \mathbf{h}_1 & \vdots & \mathbf{h}_2 & \vdots & \mathbf{h}_3 \end{bmatrix} = \begin{bmatrix} 0 & 0.64 & -0.76 \\ 1 & -0.76 & 0 \\ 0 & 0 & -0.64 \end{bmatrix} \tag{2.43}
$$

and

$$
^S\mathbf{p}_{SH} = \begin{bmatrix} 17 & 0.5 & -3 \end{bmatrix}^T \tag{2.44}
$$

As the method for determining the configuration of a rigid body with respect to a second coordinate frame is now well established through this systematic approach, it is only natural to extend this method for use in human modeling, for the purpose of simulating human motion to perform tasks in the virtual world. However, logistics regarding the embedding of coordinate frames in each rigid body (called a link) are to be developed, particularly when many segmental links and joints are needed. Similarly, a systematic representation of a homogeneous transformation matrix between two consecutive links should also be developed because coordinate frames are typically arbitrarily oriented.

2.7 Determining the position of a multi-segmental link: forward kinematics

The forward kinematics problem is characterized by determining the final position and orientation of a link (e.g., anatomical landmark on the hand) with knowledge of the joint variables. One can think of the forward kinematics (sometimes called direct kinematics) as a black box that contains the necessary calculations for accepting joint coordinates as input, and producing position and orientation parameters as output, for any of the segmental links in the chain. This black box approach is depicted in Figure 2.13.

Consider the arm of a person constrained to move on the surface of a table. Assume that this arm is represented by only two joints, characterized by two variables, q_1 and q_2. The question that will be addressed throughout this chapter is as follows:

Given a displacement of $q_1 = 15°$ and $q_2 = 30°$, what is the final position of the hand?

In order to answer this question, it is necessary to formulate an equation that contains the two independent variables q_1 and q_2 as its parameters, and that evaluates to a position. Because two parameters are involved, we would also need two independent coordinate systems, which we shall denote by frame 0 and frame 1 shown in Figure 2.14.

For this simple example, it can be seen from geometry that the x- and y-values of the hand with respect to the first coordinate system x_0 and y_0 are

$$x = 4\cos q_1 + 6\cos(q_1 + q_2) \tag{2.45}$$

and

$$y = 4\sin q_1 + 6\sin(q_1 + q_2) \tag{2.46}$$

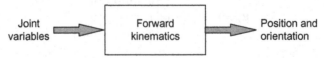

FIGURE 2.13

Understanding forward kinematics.

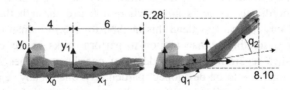

FIGURE 2.14

The upper limb with two embedded coordinate systems.

Therefore, to calculate the final position of the hand, we substitute for values of q_1 and q_2, which yields the final position of the hand as

$$\mathbf{x} = \begin{bmatrix} 8.10 & 5.28 \end{bmatrix}^T \qquad (2.47)$$

From this simple example, it can be seen that if the number of DOF becomes large, and the orientation of each joint with respect to another is spatial rather than planar, the formulation of the x, y, and z equations becomes complicated. If the orientation of the hand is required, further complexity is introduced. Therefore, we need to develop a systematic methodology for:

a. Locating coordinate systems on each segmental link in a consistent manner.
b. Calculating the relation between any two segmental links.
c. Characterizing the position and orientation of a distal link on the kinematic chain with respect to another link on the same or different chain.

2.8 The Denavit–Hartenberg representation

In order to obtain a systematic method for describing the configuration (position and orientation) of each pair of consecutive segmental links, a method was proposed by Denavit and Hartenberg (1955). We shall utilize the method of Denavit and Hartenberg (DH) to address human kinematics.

The method, now referred to as the DH method, is based upon characterizing the configuration of link i with respect to link $i-1$ by a (4×4) homogeneous transformation matrix representing each link's coordinate system. If each pair of consecutive links represented by their associated coordinate system (Figure 2.15) is related via a matrix, then using the matrix chain-rule multiplication, it is possible to relate any of the segmental links (e.g., the hand) with respect to any other segmental link (e.g., the shoulder).

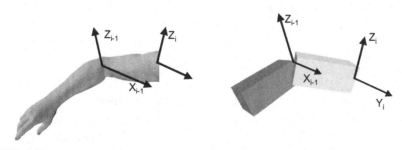

FIGURE 2.15

Joint coordinate systems between two segmental links.

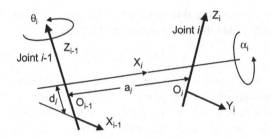

FIGURE 2.16

The relation between two coordinate systems with four parameters.

We shall always refer to **T** as having the following vectors

$$
\mathbf{T} = \begin{bmatrix} n_x & s_x & a_x & p_x \\ n_y & s_y & a_y & p_y \\ n_z & s_z & a_z & p_z \\ 0 & 0 & 0 & 1 \end{bmatrix} = \left[\begin{array}{c|c|c|c} \mathbf{n} & \mathbf{s} & \mathbf{a} & \mathbf{p} \\ \hline 0 & 0 & 0 & 1 \end{array} \right]
\tag{2.48}
$$

A single general transformation matrix between any two coordinate systems is needed. A general motion (translation and rotation) between any two configurations can be identified by four consecutive transformations taking the rigid body from one configuration to its final destination. These transformations are illustrated in Figure 2.16 and are listed as follows:

1. A rotation of α angle about the OX axis
2. A translation of a units along the OX axis
3. A translation of d units along the OZ axis
4. A rotation of θ angle about the OZ axis.

Since all transformations are about the world coordinate system, the transformation matrices are pre-multiplied. Remember that a basic homogeneous transformation matrix characterizes each transformation. The final transformation can be written as

$$
\mathbf{T}(d, \theta, \alpha, a) = \mathbf{T}_{z,\theta}\mathbf{T}_{z,d}\mathbf{T}_{x,a}\mathbf{T}_{x,\alpha} = \begin{bmatrix} \cos\theta & -\sin\theta & 0 & 0 \\ \sin\theta & \cos\theta & 0 & 0 \\ 0 & 0 & 1 & 0 \\ 0 & 0 & 0 & 1 \end{bmatrix} \begin{bmatrix} 1 & 0 & 0 & a \\ 0 & 1 & 0 & 0 \\ 0 & 0 & 1 & 0 \\ 0 & 0 & 0 & 1 \end{bmatrix}
$$

$$
\begin{bmatrix} 1 & 0 & 0 & 0 \\ 0 & 1 & 0 & 0 \\ 0 & 0 & 1 & d \\ 0 & 0 & 0 & 1 \end{bmatrix} \begin{bmatrix} 1 & 0 & 0 & 0 \\ 0 & \cos\alpha & -\sin\alpha & 0 \\ 0 & \sin\alpha & \cos\alpha & 0 \\ 0 & 0 & 0 & 1 \end{bmatrix}
$$

$$
\mathbf{T}(d, \theta, \alpha, a) = \begin{bmatrix} \cos\theta & -\cos\alpha\sin\theta & \sin\alpha\sin\theta & a\cos\theta \\ \sin\theta & \cos\alpha\cos\theta & -\sin\alpha\cos\theta & a\sin\theta \\ 0 & \sin\alpha & \cos\alpha & d \\ 0 & 0 & 0 & 1 \end{bmatrix}
$$

$$
\tag{2.49}
$$

The resulting homogeneous transformation matrix of Equation (2.49) is an important element in developing the DH representation. For any two rigid bodies, this transformation matrix characterizes the configuration (position and orientation) of one with respect to the other in terms of the four important parameters (d,θ,α,a). Therefore, if it is possible to expand this result to the representation of rigid bodies connected in a serial chain, their multiplication will yield a transformation matrix relating any two links in the chain. We will expand on this issue in the following section toward establishing a systematic method for the embedding of coordinate systems.

2.9 The kinematic skeleton

In order to establish a systematic method for biomechanically modeling human anatomy, it is necessary to establish a convention for representing segmental links and joints. We can represent human anatomy as a sequence of rigid bodies (*links*) connected by *joints*. Of course, this serial linkage could be an arm, a leg, a finger, a wrist, or any other functional mechanism. Joints in the human body vary in shape, function, and form. The complexity offered by each joint must also be modeled, to the extent possible, to enable a correct simulation of the motion. The degree by which a model replicates the actual physical model is called the level of *fidelity*.

Perhaps the most important element of a joint is its function, which may vary according to the joint's location and physiology. The physiology becomes important when we discuss the loading conditions of a joint. In terms of kinematics, we shall address the function in terms of the number of DOF associated with its overall movement. Muscle action, ligament, and tendon attachments at a joint are also important and contribute to the function.

For example, consider the elbow joint, which is considered a hinge or one-DOF rotational joint (e.g., the hinge of a door) because it allows for flexibility and extension in the sagittal plane as the radius and ulna rotate about the humerus. We shall represent this joint by a cylinder that rotates about one axis and has no other motions (i.e., one DOF). Therefore, we can now say that the elbow is characterized by one DOF and is represented throughout the book as a cylindrical rotational joint, also shown in Figure 2.17A.

On the other hand, consider the shoulder complex. The glenohumeral joint (shoulder joint) is a multi-axial (ball and socket) synovial joint between the head of the humerus and the glenoid cavity. There is a 4 to 1 incongruence between the large round head of the humerus and the shallow glenoid cavity. A ring of fibrocartilage attaches to the margin of the glenoid cavity forming the glenoid labrum. This serves to form a slightly deeper glenoid fossa for articulation with the head of the humerus Figure 2.17B.

There are a number of methods that can be used to model this complex joint (Figure 2.18). One such method (Maurel et al., 1996) is to consider the shoulder girdle (considering bones in pairs) as four joints that can be distinguished as: the sterno-clavicular joint, which articulates the clavicle by its proximal end onto the

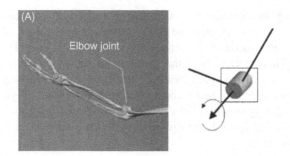

FIGURE 2.17A

A one-DOF elbow joint.

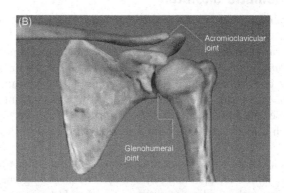

FIGURE 2.17B

The shoulder complex.

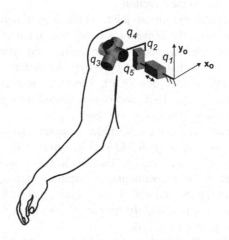

FIGURE 2.18

Modeling of the shoulder complex as three revolute and two prismatic DOFs.

sternum; the acromio-clavicular joint, which articulates the scapula by its acromion onto the distal end of the clavicle; the scapulo-thoracic joint, which allows the scapula to glide on the thorax; and the gleno-humeral joint, which allows the humeral head to rotate in the glenoid fossa of the scapula.

Another method takes into consideration the final gross movement of the joint (Abdel-Malek et al., 2001a–d), as abduction/adduction (about the anteroposterior axis of the shoulder joint), flexion/extension and transverse flexion/extension (about the mediolateral axis of the shoulder joint). Note that these motions provide for three rotational degrees of freedom having their axis intersecting at one point. This gives rise to the effect of a spherical joint typically associated with the shoulder joint (Figure 2.18). In addition, the upward/downward rotation of the scapula gives rise to two substantial translational degrees of freedom in the shoulder complex. This sliding motion is represented by elevation/depression, protraction/retraction, tipping forward/backward and medial/lateral rotations for the scapulo-thoracic joint (5 DOF).

This model allows for consideration of the coupling between some of the joints as is the case in the shoulder where muscles extend over more than one segment. When muscles are used to lift the arm in a rotational motion, unwittingly, a translational motion of the shoulder occurs.

The hand is composed of many small bones called carpals, metacarpals, and phalanges. The two bones of the lower arm—the radius and the ulna—meet at the hand to form the wrist. We will model the wrist as three intersecting revolute joints intersecting at one point, whose action yields a spherical wrist (Pieper, 1968). The complete 9-DOF model of the upper extremity is shown in Figure 2.19.

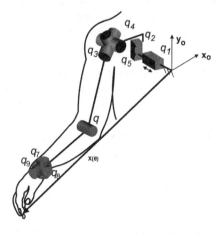

FIGURE 2.19

Modeling of the upper extremity as a 9-DOF kinematic chain.

We shall use this method to develop a complete model of the human, always maintaining a serial link approach to facilitate a computational approach. As will be seen in later chapters, this approach enables us to address problems in simulation and ergonomics.

2.10 Establishing coordinate systems

In order to obtain a systematic method for generating the (4×4) homogeneous transformation matrix between any two links, it is necessary to follow a convention in establishing coordinate systems on each link. This can be accomplished by implementing the following rules. It should be emphasized that a suitable *home configuration* must first be established before applying these rules. A home configuration denotes the start configuration of the serial chain (segmental links). It is customary to start from a well-known position where the user indicates that this posture is the home configuration.

The procedure for establishing coordinate frames at each link is as follows:

1. Name each joint starting with $1,2,\ldots$ up to n-degrees of freedom.
2. Embed the $\mathbf{z}_{i-1}$ axis along the axis of motion of the ith joint.
3. Embed the $\mathbf{x}_i$ axis normal to the $\mathbf{z}_{i-1}$ (and of course normal to the $\mathbf{z}_i$ axis).
4. Embed the $\mathbf{y}_i$ axis such that it is perpendicular to the $\mathbf{x}_i$ and $\mathbf{z}_i$ subject to the right hand rule. However, on the kinematic skeleton, it is customary not to show the $\mathbf{y}_i$ axis so as not to clutter the drawing and since it is not needed for determining the DH parameters.

The location of the origin of the first coordinate frame (frame 0) can be chosen to be anywhere along the $\mathbf{z}_0$ axis. In addition, for the nth coordinate system, it can be chosen to be embedded anywhere in the nth link subject to the above four rules. In order to generate the matrix relating any two links, four parameters are needed. The four parameters are:

1. θ_i is the joint angle, measured from the $\mathbf{x}_{i-1}$ to the $\mathbf{x}_i$ axis about the $\mathbf{z}_{i-1}$ (right hand rule applies). For a prismatic joint θ_i is a constant. It is basically the angle rotation of one link with respect to another about the $\mathbf{z}_{i-1}$.
2. d_i is the distance from the origin of the $(i-1)^{\text{th}}$ coordinate frame to the intersection of the $\mathbf{z}_{i-1}$ axis with the $\mathbf{x}_i$ axis along $\mathbf{z}_{i-1}$ axis. For a revolute joint, d_i is a constant. It is basically the distance translated by one link with respect to another along the $\mathbf{z}_{i-1}$ axis.
3. a_i is the offset distance from the intersection of the $\mathbf{z}_{i-1}$ axis with the $\mathbf{x}_i$ axis to the origin of the ith frame along $\mathbf{x}_i$ axis. (Shortest distance between the $\mathbf{z}_{i-1}$ and $\mathbf{z}_i$ axis).
4. α_i offset angle from $\mathbf{z}_{i-1}$ axis to $\mathbf{z}_i$ axis about the $\mathbf{x}_i$ axis (right hand rule).

Careful attention must be given when the following cases occur:

1. When two consecutive axes are parallel, the common normal between them is not uniquely defined, i.e., the direction of x_i must be perpendicular to both axes; however, the position of x_i is arbitrary.
2. When two consecutive axes intersect, the direction of x_i is arbitrary.

The four values for the DH parameters $\theta_i, d_i, a_i, \alpha_i$ are typically entered into a table known as the DH table. A 10-DOF model will have a table with 10 rows. Each row is used to generate the homogeneous transformation matrix for a DOF.

2.10.1 Example: a 9-DOF model of an upper limb

A 9-DOF model of the upper extremity is shown in Figure 2.20. Sketch the coordinate systems for each joint according to the DH representation method (note that the first two joints are prismatic (translational).

Each z-axis, starting with the first joint and denoted by z_0, is located along the joint (Figure 2.21). The x-axis is also located but the y-axis is not shown to avoid cluttering the figure.

The position and location of each axis determines the parameters $\theta_i, d_i, a_i, \alpha_i$, and hence determine the resulting (4×4) homogeneous transformation matrix from Equation (2.49), and repeated here for continuity.

$$
{}^{i-1}\mathbf{T}_i = \begin{bmatrix} \cos\theta_i & -\cos\alpha_i\sin\theta_i & \sin\alpha_i\sin\theta_i & a_i\cos\theta_i \\ \sin\theta_i & \cos\alpha_i\cos\theta_i & -\sin\alpha_i\cos\theta_i & a_i\sin\theta_i \\ 0 & \sin\alpha_i & \cos\alpha_i & d_i \\ 0 & 0 & 0 & 1 \end{bmatrix} \tag{2.50}
$$

For any sequence of consecutive transformations or for any serial mechanism, any two reference frames can be represented by the multiplication of the transformations between them. To relate the coordinates frames $m-1$ and $m+n$, the following transformations are computed

$$
{}^{m-1}\mathbf{T}_m \, {}^{m}\mathbf{T}_{m+1} \cdots \cdots {}^{m+n-1}\mathbf{T}_{m+n} = {}^{m-1}\mathbf{T}_{m+n} \tag{2.51}
$$

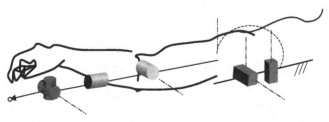

FIGURE 2.20

A 9-DOF model of the upper extremity.

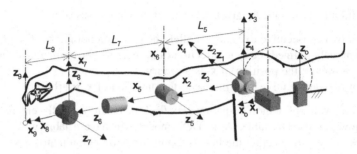

FIGURE 2.21

Locating the coordinate systems for the 9-DOF model of the upper extremity.

In general, the kinematic equations of serial n-DOF segmental links relating the coordinates of the last link (the nth coordinate frame) with respect to the world coordinates system (the 0th coordinate frame) can be written as

$$^0T_i = {^0T_1}{^1T_2}...{^{i-1}T_i} = \prod_{j=1}^{i} {^{j-1}T_j} \text{ for } i = 1, 2, ..., n \tag{2.52}$$

For example, for a 9-DOF model of the upper limb ($n = 9$), the position and orientation of the 9th link with respect to the base frame (the first frame) is represented by:

$$^0T_1 {^1T_2}...{^8T_9} = {^0T_9} \tag{2.53}$$

If ^{m-1}x and ^mx are the extended position vectors of a point, referred to coordinate frames embedded in link $m-1$ and m, respectively, the relationship between the two vectors is given by

$$^{m-1}x = {^{m-1}T_m}{^mx} \tag{2.54}$$

Similarly, a vector resolved in the coordinates of the hand (^nv) can be resolved in the world coordinate system by multiplying by the corresponding transformation matrix as

$$^0v = {^0T_n}{^nv} \tag{2.55}$$

where the vector ^0v is resolved in the world coordinate frame.

2.10.2 Example: DH parameters of the lower limb

Consider the lower limb shown in Figure 2.22. For the purpose of this example, a total of four DOF are used to model the limb.

a. Determine the coordinate systems and sketch on the figure
b. Determine the DH table
c. Write down the matrices relating any two consecutive coordinate systems
d. Write down the matrix relating the first link to the coordinate system embedded in the foot.

FIGURE 2.22

A simple 4-DOF model of the lower limb.

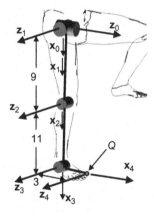

FIGURE 2.23

Establishing coordinate systems on a 4-DOF model of the lower limb.

e. If the lower limb has moved to $q_1 = 0$, $q_2 = 90^o$, $q_3 = -90^o$, and $q_4 = 0$, calculate the final orientation of the foot, and the final coordinates of the point Q located at the tip of the foot.

Results for (a) to (e) above:

a. Coordinate systems are established as shown in Figure 2.23.
b. The DH parameters are shown in Table 2.1.
c. Substituting each row of the DH table into Equation (2.50) yields the four transformation matrices as follows:

$$
{}^0\mathbf{T}_1 = \begin{bmatrix} \cos q_1 & 0 & \sin q_1 & 0 \\ \sin q_1 & 0 & -\cos q_1 & 0 \\ 0 & 1 & 0 & 0 \\ 0 & 0 & 0 & 1 \end{bmatrix} \tag{2.56}
$$

Table 2.1 DH Parameters for the Lower Limb

Joint	θ_i	α_i	a_i	d_i
1	q_1	0	0	$\pi/2$
2	q_2	0	9	0
3	q_3	0	11	0
4	$q_4 + \pi/2$	0	3	0

$$^1\mathbf{T}_2 = \begin{bmatrix} \cos q_2 & -\sin q_2 & 0 & 9\cos q_2 \\ \sin q_2 & \cos q_2 & 0 & 9\sin q_2 \\ 0 & 0 & 1 & 0 \\ 0 & 0 & 0 & 1 \end{bmatrix} \tag{2.57}$$

$$^2\mathbf{T}_3 = \begin{bmatrix} \cos q_3 & -\sin q_3 & 0 & 11\cos q_3 \\ \sin q_3 & \cos q_3 & 0 & 11\sin q_3 \\ 0 & 0 & 1 & 0 \\ 0 & 0 & 0 & 1 \end{bmatrix} \tag{2.58}$$

$$^3\mathbf{T}_4 = \begin{bmatrix} -\sin q_4 & -\cos q_4 & 0 & -3\sin q_4 \\ \cos q_4 & -\sin q_4 & 0 & 3\cos q_4 \\ 0 & 0 & 1 & 0 \\ 0 & 0 & 0 & 1 \end{bmatrix} \tag{2.59}$$

d. To relate the first frame to the fourth frame, the following multiplication of homogeneous transformation matrices must be carried out:

$$^0\mathbf{T}_1 \, ^1\mathbf{T}_2 \, ^2\mathbf{T}_3 \, ^3\mathbf{T}_4 = \, ^0\mathbf{T}_4 \tag{2.60}$$

where the resulting (4×4) matrix

$$^0\mathbf{T}_4 = \begin{bmatrix} n_x & o_x & a_x & p_x \\ n_y & o_y & a_y & p_y \\ n_z & o_z & a_z & p_z \\ 0 & 0 & 0 & 1 \end{bmatrix} \tag{2.61}$$

where,

$$n_x = -\cos q_1 \sin(q_2 + q_3 + q_4) \tag{2.62}$$

$$n_y = -\sin q_1 \sin(q_2 + q_3 + q_4) \tag{2.63}$$

$$n_z = \cos(q_2 + q_3 + q_4) \tag{2.64}$$

$$s_x = -\cos q_1 \cos(q_2 + q_3 + q_4) \tag{2.65}$$

$$s_y = -\sin q_1 \cos(q_2 + q_3 + q_4) \tag{2.66}$$

$$s_z = -\sin(q_2 + q_3 + q_4) \tag{2.67}$$

$$a_x = \sin q_1 \tag{2.68}$$

$$a_y = -\cos q_1 \tag{2.69}$$

$$a_z = 0 \tag{2.70}$$

$$p_x = \cos q_1(-3\sin(q_2 + q_3 + q_4) + 9\cos q_2 + 11\cos(q_2 + q_3)) \tag{2.71}$$

$$p_y = \sin q_1(-3\sin(q_2 + q_3 + q_4) + 9\cos q_2 + 11\cos(q_2 + q_3)) \tag{2.72}$$

$$p_z = 9\sin q_2 + 11\sin(q_2 + q_3) + 3\cos(q_2 + q_3 + q_4) \tag{2.73}$$

e. To determine the orientation of the 4th coordinate system with respect to the hip, we substitute $q_1 = 0$, $q_2 = 0$, $q_3 = 0$, and $q_4 = 0$ into $^0\mathbf{T}_4$ and the orientation is identified as $\mathbf{n} = \begin{bmatrix} 0 & 0 & 1 \end{bmatrix}^T$, $\mathbf{s} = \begin{bmatrix} -1 & 0 & 0 \end{bmatrix}^T$, and $\mathbf{a} = \begin{bmatrix} 0 & -1 & 0 \end{bmatrix}^T$.

In order to determine the position of the point Q, defined with respect to the 4th coordinate system, we shall use the extended vector equation as

$$\begin{bmatrix} ^0\mathbf{v} \\ 1 \end{bmatrix} = {}^0\mathbf{T}_4(\mathbf{q}) \begin{bmatrix} ^4\mathbf{v} \\ 1 \end{bmatrix} \tag{2.74}$$

The position of a point on the foot is given by $\mathbf{v}_Q = \begin{bmatrix} 0 & 0 & 0 \end{bmatrix}^T$. For the initial posture of the lower limb, the joints are $q_1 = 0$, $q_2 = 0$, $q_3 = 0$, and $q_4 = 0$.

In order to calculate the new posture given the change in joint variables, we substitute into $^0\mathbf{T}_4$

$${}^0\mathbf{T}_4(0, 90°, -90°, 0) = \begin{bmatrix} 0 & -1 & 0 & 11 \\ 0 & 0 & -1 & 0 \\ 1 & 0 & 0 & 12 \\ 0 & 0 & 0 & 1 \end{bmatrix} \tag{2.75}$$

The position is calculated as

$$\mathbf{v}_Q(0, 90°, -90°, 0) = \begin{bmatrix} 12 & 0 & -11 \end{bmatrix} \tag{2.76}$$

and the orientation is calculated as

$$\mathbf{n} = \begin{bmatrix} 1 & 0 & 0 \end{bmatrix}^T \tag{2.77}$$

$$\mathbf{s} = \begin{bmatrix} 0 & 0 & 1 \end{bmatrix}^T \tag{2.78}$$

$$\mathbf{a} = \begin{bmatrix} 0 & -1 & 0 \end{bmatrix}^T \tag{2.79}$$

as shown in Figure 2.24.

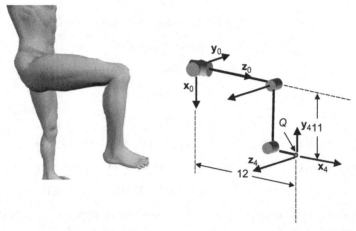

FIGURE 2.24

The configuration of the lower limb as a 4-DOF model.

2.11 The Santos® model

Figure 2.25 illustrates Santos® with the kinematic model (skeleton). The values L_i between joints define the body parameters of the human, such as shoulder-to-elbow distance. For generality of use of the human model, these parameters are left as user-input and can be changed to correspond to specific anthropometrical data. The high level of control over the distances L_i allows this model to represent any anthropometrical percentile.

The z-axes in Figure 2.25 represent the axes of rotation for each DOF (each kinematic revolute joint). The model includes four 3-DOF spherical joints (each modeled using three revolute joints) to represent movement of the human spine, totaling 12 DOF for the torso segment (labeled z_0 through z_{11}). Each arm includes two revolute joints for the clavicle, a spherical joint for the shoulder, and four additional DOF for the elbow and wrist. Hence, each arm has 9 DOF (labeled z'_{12} through z'_{20} for the right arm and z_{12} through z_{20} for the left). Finally, the leg segments have 7 DOF each, including a spherical joint at each hip. The result is a segmented human kinematic model capable of representing human motion with considerable accuracy.

2.12 Variations in anthropometry

Because the DH method is dependent upon a well-defined set of parameters, as entered into the DH Table, it is indeed straightforward to vary these parameters as the anthropometric model is changed.

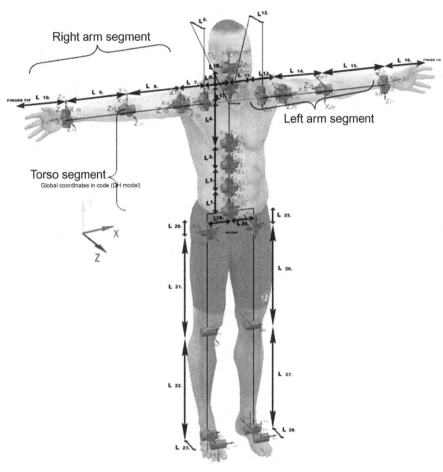

Right arm segment

Left arm segment

Torso segment
Global coordinates in code (DH model)

FIGURE 2.25

Santos®—a 215-DOF human model.

2.13 A 55-DOF whole body model

A kinematic human skeletal model with 55 DOFs, shown in Figure 2.26, consists of six physical branches and one virtual branch. The physical branches include the right leg, the left leg, the spine, the right arm, the left arm, and the head. In these branches, the right leg, the left leg, and the spine start from the pelvis (z_4, z_5, z_6), while the right arm, left arm and head start from the spine end joint (z_{30}, z_{31}, z_{32}). This model shall be used in predictive dynamics to account for the physical translational and rotational joints of the human with respect to a fixed coordinate system on the ground.

The spine model includes four joints, each joint has three rotational DOFs ([z_{21}, z_{22}, z_{23}], [z_{24}, z_{25}, z_{26}], [z_{27}, z_{28}, z_{29}], [z_{30}, z_{31}, z_{32}]). The legs and arms are

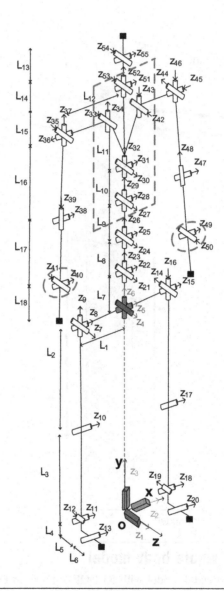

FIGURE 2.26

The 55-DOF digital human model (with global DOFs z_1, z_2, z_3, z_4, z_5, z_6) used in predictive dynamics.

assumed to be symmetric along the sagittal plane. Each leg consists of a thigh, a shank, a rear foot, and a forefoot. There are seven DOFs for each leg: three at the hip joint (z_7, z_8, z_9), one at the knee joint (z_{10}), two at the ankle joint (z_{11}, z_{12}), and finally, one to characterize the forefoot (z_{13}). At the clavicle, there are two orthogonal revolute joints (z_{33}, z_{34}). Each arm consists of an upper arm, a lower arm, and a hand. There are seven DOFs for each arm: three at the shoulder, two at the elbow, and two at the wrist. In addition, there are five DOFs for the head

Table 2.2 Link Length and Mass Properties

Link	Length (cm)	Mass (kg)
L_1	8.51	4.48
L_2	38.26	9.54
L_3	39.46	3.74
L_4	5.0	0.5
L_5	9.01	0.7
L_6	7.56	0.23
L_7	9.0	2.32
L_8	5.63	2.32
L_9	5.44	2.32
L_{10}	6.0	2.32
L_{11}	17.39	3.0
L_{12}	16.76	5.78
L_{13}	20.0	4.22
L_{14}	17.1	1.03
L_{15}	4.41	2.8
L_{16}	25.86	1.9
L_{17}	24.74	1.34
L_{18}	16.51	0.5

branch: three at the lower neck and two at the upper neck. The anthropometric data for the skeletal model representing a 50-percentile male, generated using commercial software, are shown in Table 2.2.

It is noted here that the foregoing 55-DOF skeletal model has been developed to simulate many human activities, such as symmetric and asymmetric walking, running, stairs climbing, lifting objects, throwing, etc. For simulating each of these activities, some DOF that do not participate in the activity in a significant manner are frozen to their neutral angles. The formulation presented here is quite flexible allowing any DOF to be frozen to a specified value. In addition, limits on the range of motion of any DOF can be imposed. A general-purpose software is being developed that can be used to simulate these activities using the same skeletal model. For the symmetric gait simulation problem, the following DOFs are frozen: wrist joint, clavicle joint, neck joint, and two spine joints (shown as red dashed enclosures in Figure 2.26). Therefore, the skeletal model used for gait simulation has 38 active DOFs. The other way to accomplish this objective would be to redefine the skeletal model for each activity. However, this would require redefinition of the body segments and recalculation of their mass and inertial properties, which would be quite tedious.

2.14 Global DOFs and virtual joints

The six global DOFs generate rigid body motion for the entire spatial skeleton model. The three translations are represented by three prismatic joints, and the three rotations by three revolute joints in the DH method. These joints are named as virtual joints to distinguish them from the physical human joints. The two adjacent virtual joints are connected by a virtual link which uses zero mass and zero inertia to define the link properties. Finally, the virtual joints and links constitute a virtual branch which contains six global DOFs (z_1, z_2, z_3, z_4, z_5, z_6).

The virtual joints defined in the virtual branch not only generate global rigid body movements but also contain global generalized forces. These forces correspond to the six global DOFs: three forces (τ_1, τ_2, τ_3) and three moments (τ_4, τ_5, τ_6). For the system in equilibrium, these global generalized forces should be zero.

2.15 Concluding remarks

This chapter has presented the kinematics of human modeling. It is evident that the human body is complex, requiring a true multi-disciplinary approach among researchers from the medical field and from engineering. Detailed physics-based modeling of human joints may require far more knowledge than is currently available. However, for all practical purposes, it has been shown that approximate modeling of gross human motion, for the purpose of human motion simulation or ergonomic analysis, can be achieved using the Denavit–Hartenberg representation method. The DH method provides an adequate, consistent, and systematic method for embedding the local coordinate systems for each link. Furthermore, the DH method allows for the modeling of complex and large DOF skeletons, where each complex joint has multiple DOF. Lastly, the DH method lends itself well to computational methods and we will expand on this foundation in later chapters to build the dynamics of the motion, leading to predicting how humans will respond to a given situation.

References

Abdel-Malek, K., Yu, W., Jaber, M., 2001a. Realistic Posture Prediction. 2001 SAE Digital Human Modeling and Simulation.

Abdel-Malek, K., Wei, Y., Mi, Z., Tanbour, E., Jaber, M., 2001b. Posture prediction versus inverse kinematics. In: Proceedings of the 2001 ASME Design Engineering Technical Conferences and Computers and Information in Engineering Conference. Pittsburgh, PA, pp. 37–45.

Abdel-Malek, K., Yang, J., Brand, R., Tanbour, E., 2001c. Towards understanding the workspace of the upper extremities. SAE Trans. J. Passenger Cars: Mech. Syst. 110 (6), 2198–2206.

Abdel-Malek, K., Yu, W., Jaber, M., Duncan, J., 2001d. Realistic Posture Prediction for Maximum Dexterity. SAE Technical Paper 2001-01-2110. doi:10.4271/2001-01-2110.

Denavit, J., Hartenberg, R.S., 1955. A kinematic notation for lower-pair mechanisms based on matrices. J. Appl. Mech. 77, 215–221.

Maurel, W., Thalmann, D., Hoffmeyer, P., Beylot, P., Gingins, P., Kalra, P., et al., 1996. A biomechanical musculoskeletal model of human upper limb for dynamic simulation. In: Computer Animation and Simulation. Springer Vienna, pp. 121–136.

Pieper, D.L., 1968. The Kinematics of Manipulators Under Computer Control. Ph.D. Thesis, Stanford University.

Posture Prediction and Optimization

The concept is interesting and well-formed, but in order to earn better than a 'C', the idea must be feasible.

A Yale University management professor in response to student Fred Smith's paper proposing reliable overnight delivery service (Smith went on to found Federal Express Corp.)

3.1 What is optimization?

Mathematical optimization is a branch of computational mathematics that seeks to answer the question: "What is the best solution given a set of constraints?"

This type of optimization is well suited to problems for which the quality of any answer can be expressed as a numerical value. Such problems arise in many fields, particularly in business, engineering, structures, architecture, economics, management, and just about every smart appliance on the market today. Equally wide in scope is the range of methods used to solve such problems. An excellent reference on the subject, including solution methods, is a text written by the second author (Arora, 2012).

This chapter presents the basic ideas of mathematical optimization and provides a rigorous understanding of the three ingredients: design variables, cost functions, and constraints. The chapter does not, however, cover numerical methods or numerical solvers of optimization as this is quite a mature field and there are many existing computational codes that can be used to solve a well-formulated optimization problem. It is believed that optimal solutions are readily obtained if the problem is well formulated.

3.2 What is posture prediction?

Posture prediction refers to the estimation of joint variables that will allow the human body to assume a posture towards achieving an objective. For example, prediction of upper extremity variables (joint angles) to achieve the grasping of an object is a posture prediction problem. Note that grasping means the positioning of the hand (called the end-effector of a serial kinematic chain in the field of robotics), often referred as inverse kinematics or IK for short. Similarly, calculating the joint variables of a lower extremity to allow the foot to be positioned on a pedal, for example, is also a posture prediction problem.

For a posture prediction problem, the design variables are the joint angles of the body. The constraints are the location and possibly the orientation of an end-effector (usually the fingertip) and the joint ranges of motion. Commonly, optimization-based inverse kinematic methods focus on minimizing some form of discomfort (Abdel-Malek et al., 2004b; Jung and Choe, 1996) or minimizing perturbation from a neutral position (Case et al., 1990; Porter et al., 1990). If more than one objective function is used, they can be combined using multi-objective optimization techniques like the objective sum method, the min−max method, or the global criterion method (Yang et al., 2004a−d). Other objective functions that have been tried include minimizing joint displacement, minimizing change in potential energy, minimizing the distance to the target, and maximizing reachability (Abdel-Malek et al., 2004a,c; Yang et al., 2004a−d).

Optimization can be used not only to find the most realistic posture to reach a point in the workspace, but also to find the best placement of the human within a workspace or the best placement of the target relative to the human (Abdel-Malek et al., 2004a,c; Abdel-Malek et al., 2006; Abdel-Malek et al., 2001a−d; Abdel-Malek et al., 2005; Marler et al., 2009; Mi et al., 2002a; Yang et al., 2006a,b,c,d; Mi et al., 2009; Yang et al., 2004a−d).

Motion prediction broadens the approach and finds the optimal motion between two target points. For motion prediction, B-spline approximations or polynomials are used to approximate joint displacement with respect to time (Mi, 2004). Furthermore, additional objective functions are formulated that deal with the change in joint displacement over time. Examples include minimizing inconsistency, minimizing joint acceleration, and minimizing velocity at the beginning and end of the motion (Abdel-Malek et al., 2004b; Mi, 2004).

In this chapter, the treatment of an open kinematic chain is addressed as opposed to closed-loop systems. For the human body, we consider a variety of open kinematic chains, typically all beginning in the waist and extending to the hand, the foot, or the head. Similarly, the kinematic chain can begin with the ground (foot); then extend through the body to the arm, hand, and fingers. Note that the word *prediction* is often used rather than *calculation* since predicting realistic human motion is not an exact science, but rather a prediction of human behavior. For a human to touch a given point with a finger, several postures exist…but what is the "best" way to accomplish this task?

Motion prediction refers to the calculation of joint variables with respect to time (often called a motion profile) to enable motion from an initial configuration to a final configuration. Posture prediction is typically associated with the calculation of static postures, which characterizes the final positions and orientations of segmental links without much consideration of how this posture occurred. The challenge in predicting postures and motions is evident in that there are a large (infinite) number of solutions.

There are many theories and methodologies for how the brain issues motor control commands to the central nervous system to achieve a task. This chapter will not address all reported methods but rather will focus on the method that is based on optimization, which we believe is a natural process for completing a task.

3.3 Inducing behavior

One method for characterizing brain function to induce behavior in a digital human is to give it the ability to process a task using the previously mentioned cost functions. This theory was first introduced by the first author and his colleagues some time ago (Abdel-Malek et al., 2001a–d). A task is first broken into sub-tasks, and then each sub-task is broken into procedures. This process is called task planning and is well understood in the field of robotics. Subsequently, to accomplish a procedure, the procedure planner selects one or more cost functions from a list of available human performance measures to be used in an optimization algorithm (Figure 3.1). Note that task planning and procedure planning are forms of intelligent engines. Indeed, it is believed that selecting the appropriate type of cost functions and understanding their relative importance is a challenging problem that is not well understood. Nevertheless, optimizing for a combination of cost functions yields a behavior that is different than that yielded by optimizing

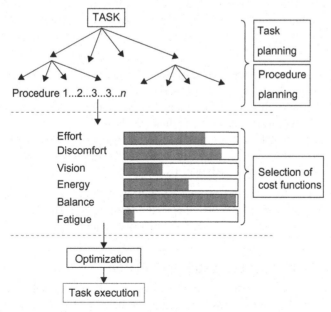

FIGURE 3.1

The overall planning and execution of a task using the theory of optimization.

for a different set of cost functions. It is expected that this aspect of selecting cost functions will continue to be an area of research of great interest as it produces behaviors that vary for performing the same task with different cost functions.

The main new idea here is that humans execute procedures while minimizing or maximizing cost functions. By selecting a combination of various cost functions, a behavior is induced and a new motion profile is calculated. This concept will be explained further in this chapter.

3.4 Posture prediction versus inverse kinematics

Because of the apparent mathematical complexity of the problem of posture prediction and because the human body has many more than six DOF, methods for predicting postures are quite involved. Posture prediction, at least on the surface, is equivalent to what is called inverse kinematics in the field of mechanisms and robotics (known as IK in the gaming and animation industries). In this section we comment about their benefits and shortcomings.

3.4.1 Analytical and geometric IK methods

Analytical and geometric IK methods are typically associated with robotic manipulator arms that have no more than 6 DOF and where closed-form or numerical solutions are possible. Larger than 6-DOF systems become very redundant and require very complex numerical algorithms. Analytical or geometric IK methods for human posture prediction are almost impossible because of the following reasons:

1. The large number of DOF associated with the human model leads to severe difficulties in calculating solutions. It is almost impossible to use geometric methods and very difficult to use numerical methods, as identifying and finding all solutions is a difficult mathematical problem.
2. The need to calculate realistic postures. While some of the analytical methods may yield solutions, a choice of "the best" solution that looks most natural is difficult to achieve using analytical IK methods.

The main benefit of analytical and geometric methods is that, if determined, they are determined very quickly. Typically, analytical methods, even though they are numerical in nature, are computationally efficient.

3.4.2 Empirically-based posture prediction

This method is based on gathering a great deal of data from actual subjects while they are performing various postures. Anthropometry and posture data are captured and recorded. Statistical models, typically nonlinear regression, are then developed and used in posture prediction algorithms. The benefit of this method is the ability to predict postures that already have been recorded for the exact anthropometric model. Extrapolating postures and variations thereof is extremely difficult and highly inaccurate. However, where this method completely fails is in predicting motion, which is the ultimate objective of posture prediction. If it is to be expanded to predict motion, the variability and many parameters associated with motion, including dynamic effects and inertia, are not only difficult to measure but impossible to correlate. The method requires an exhaustive and often very costly experimental setup involving thousands of subjects to generate a modest model for a small population.

3.5 **Optimization-based posture prediction**

This chapter introduces a framework and associated algorithm for predicting postures that are based on an individual task. In order to better understand the motivation behind cost functions, consider the case of a driver, in a vehicle, who is about to reach for a radio control button on the dashboard. It is believed that the driver will reach directly to the button while exerting minimum effort and perhaps expending minimum energy. However, the same driver when negotiating a curve will have to place their hand on the steering wheel in such a way to be able to exert the necessary force needed to turn the wheel. As a result, involuntarily, the driver will select a posture that maximizes force at the hand, minimizes the torque at each joint, minimizes energy, and minimizes effort needed to accomplish this task (Figure 3.2).

Therefore, our underlying plot is that each task is driven by the optimization of one or more cost functions, leading the person to posture their body in the most natural manner. Note that simple logic has been implemented in the *Processor* module to correlate between the task and the cost functions.

What we have proposed through this simple example is that humans assume different postures for different tasks. Note that, when coupled with the DH

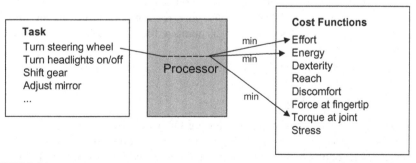

FIGURE 3.2

The task-based approach to selecting cost functions.

parameters from Chapter 2 and cost functions as human performance measures, this presents a very robust method for creating and inducing different behavior while calculating naturalistic motions.

The optimization formulation contains three main ingredients:

1. *A cost function*: A cost function is a human performance measure to be minimized or maximized. In our case, there are many cost functions, forming a multi-objective optimization problem.
2. *Design variables*: The variables are individual DOF joint displacements that will be calculated from the algorithm. In our case, these are the joint variables that define the position and orientation of each segmental link.
3. *Constraints*: Constraints are mathematical expressions that bound the problem. A constraint that makes the distance between the end-effector and the target point be within a specified tolerance is imposed. Joint ranges of motion are necessary constraints, but additional constraints are added as appropriate.

Many cost functions exist, and others are being developed by many researchers and in many fields. We will demonstrate the use of a number of such functions for posture prediction, which will then be used in the formulation for predictive dynamics in a later chapter.

This chapter focuses on introducing a general optimization-based formulation for predicting postures. These postures compare well with those assumed by real humans.

The optimum posture (set of q-values) is determined by solving the following optimization problem:

$$\text{Find:} \quad \mathbf{q} \in R^{DOF}$$
$$\text{Minimize:} \quad Discomfort, \ Effort, \text{etc.} \tag{3.1}$$

$$\text{Subject to:} \left\| \mathbf{x(q)}^{end\text{-}effector} - \mathbf{x}^{targetpoint} \right\|^2 \leq \varepsilon$$
$$\text{and} \quad q_i^L \leq q_i \leq q_i^U; \quad i = 1, 2, \ldots, DOF \tag{3.2}$$

where $\mathbf{q}$ is the vector of generalized joint variables and $\mathbf{x}^{target\ point}$ is the point in space that will be touched. The *feasible space* for a problem such as the one in Equation (3.2) is defined as the set of all points $\mathbf{x}$ for which all of the constraints are satisfied.

3.5.1 Design variables

As suggested earlier, the design variables are the generalized coordinates q_i, or in vector form, $\mathbf{q} = [q_1 \quad \ldots \quad q_{DOF}]^T$. Since all joints are rotational, q_i have units of radians. Many optimization algorithms require an *initial guess*, which entails determining initial values for the design variables. The initial guess can be somewhat arbitrary, although it is helpful to use a feasible point. In this study, an initial guess is used that satisfies the joint limits but does not necessarily satisfy the distance constraint, which is discussed below.

3.5.2 Constraints

The first constraint in Equation (3.2) is the distance constraint and requires that the end-effector contact a predetermined target point (user-specified) in Cartesian space, where ε is a small positive number that approximates zero. This constraint represents distance squared. The DH method is used to determine the position of the end-effector after a series of given displacements. In addition, each generalized coordinate is constrained to lie between lower and upper limits, represented by q_i^L and q_i^U, respectively. These limits ensure that the digital human does not assume an unrealistic position in an effort to contact the target point.

3.5.3 Cost function

For simplicity, and to demonstrate the concept, consider the first objective function, which represents a simple form of discomfort and is proportional to the deviation from the *neutral position*. The neutral position is selected as a relatively comfortable posture, typically a standing position with arms at one's sides, where q_i^N is the neutral position of a joint, and $\mathbf{q}^N$ represents the overall posture. Technically, the displacement from the neutral position is given by $|q_i - q_i^N|$; however, for computational simplicity $(q_i - q_i^N)^2$ is used. Because motion in some joints contributes more significantly to discomfort, a weighting parameter w_i is introduced to stress the importance of a particular joint. Currently, these weights are determined by trial and error. Then, the cumulative discomfort (of all joints) is characterized by the following objective function:

$$f_{Discomfort}(\mathbf{q}) = \sum_{i=1}^{n} w_i (q_i - q_i^N)^2 \tag{3.3}$$

Note that the intent in developing this performance measure is not necessarily to quantify discomfort. Rather, the function in Equation (3.3) is designed simply to yield a realistic posture and is effective in that capacity.

3.6 A 3-DOF arm example

Consider, for example, the simplified planar 3-DOF arm shown in Figure 3.3. This is an illustrative example to demonstrate our formulation where we have restricted the arm to planar motion.

The objective of this exercise is twofold:

1. Given the point $\mathbf{p} = [5.67\ 3.59]$ and the following joint ranges of motion $A = -\frac{\pi}{3} < q < \frac{\pi}{3}$, it is required to determine whether this point is reachable.

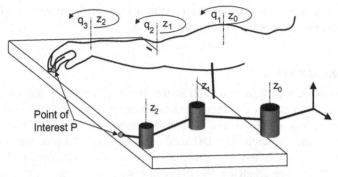

FIGURE 3.3

The upper extremity modeled as 3 DOF and restricted to planar motion.

Table 3.1 DH Table

	θ_i	d_i	α_i	a_i
1	q_1	0	0	4
2	q_2	0	0	2
3	q_3	0	0	1

2. If this point is reachable and it is required to reach the point with the orientation $\mathbf{a} = [1\ 0]$, it is necessary to calculate a posture of the upper extremity.

The DH table is readily determined and presented in Table 3.1.

Substituting each row into the general DH matrix representation, Equation (2.49) yields the following (4×4) transformation matrices

$$\mathbf{T}_{z,\theta} = \begin{bmatrix} \cos\theta & -\sin\theta & 0 & 0 \\ \sin\theta & \cos\theta & 0 & 0 \\ 0 & 0 & 1 & 0 \\ 0 & 0 & 0 & 1 \end{bmatrix} \tag{3.4}$$

Performing the multiplication and obtaining the position vector yields

$$\mathbf{x}_0^n(\mathbf{q}) = \begin{bmatrix} 4\cos q_1 + 2\cos (q_1 + q_2) + \cos (q_1 + q_2 + q_3) \\ 4\sin q_1 + 2\sin (q_1 + q_2) + \sin (q_1 + q_2 + q_3) \end{bmatrix} \tag{3.5}$$

We shall also impose ranges of motion on each joint as $-\pi/3 \le q_i \le \pi/3$; $i = 1, 2, 3$.

We now answer the two objectives:

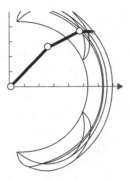

FIGURE 3.4

The computed posture is shown within the boundary of the workspace.

To predict a posture at this position **p** with the given orientation, we implement the minimum effort cost function given by:

$$f_{Effort}(\mathbf{q}) = \sum_{i=1}^{n} \gamma_i(q_i - q_i^{initial})^2 \qquad (3.6)$$

where γ_i is a weight assigned to each joint and is comparable to w_i.

The computed joint angles for the final posture are $\mathbf{q} = [0.78 \ \pi/180 \ 0.39 \ \pi/180 \ 0.39 \ \pi/180]^T$. The posture is schematically represented in Figure 3.4. The boundary of the workspace of all points in space that can be touched by the end-effector is also shown. This workspace is also called the reach envelope. Computational and closed-form methods for determining the reach envelope for humans was provided by Abdel-Malek et al. (2001a–d, 2004a–d).

3.7 Development of human performance measures

The objective functions in the final optimization formulation represent human performance measures. In this section, we first describe *joint displacement*, which is based on work by Jung and Choe (1996) and Yu (2001). We explain how this performance measure can be modified to represent *effort*. We then introduce additional performance measures.

It has been suggested that each DOF, or each body segment, should be associated with an individual objective function or component that may be considered in the final performance measure (Zacher and Bubb, 2004). Several performance measures will also be used as multi-objective optimization problems towards predicting postures.

3.7.1 Joint displacement

The neutral position $\mathbf{q}^N$ represents a relatively comfortable position. Consider q_i^N as the *neutral position* of a joint measured from the *home configuration*, which is characterized by $\mathbf{q} = \mathbf{0}$. Then, conceptually, the displacement from the neutral position for a particular joint is given by $|q_i - q_i^N|$. However, to avoid numerical difficulties and non-differentiability, the terms $(q_i - q_i^N)^2$ are used. Each of these terms (one for each DOF) serves as an individual objective function. The terms are combined using a weighted sum. The scalar weights w_i are used to stress the importance of particular joints. The consequent joint displacement function is given as follows:

$$f_{Joint\ displacement}(\mathbf{q}) = \sum_{i=1}^{DOF} w_i(q_i - q_i^N)^2 \tag{3.7}$$

We have determined the values for the weights based on trial-and-error experimentation with the 21-DOF model, and they are given in Table 3.2.

For this model, the neutral position is chosen based on observation of the skinned model in the previous chapter (see Figure 2.25) rather than a skeletal model. The resulting vector $\mathbf{q}^N$ is defined as

$$q_i^N = 0; \quad i = 1,\ldots,12,19,20$$
$$q_{13}^N = -15.0, \ q_{14}^N = 20.0, \ q_{15}^N = 100.0, \ q_{16}^N = -10.0, \ q_{17}^N = -80.0, \tag{3.8}$$
$$q_{18}^N = -35.0, \ q_{21}^N = 15.0$$

This generally represents a posture with the arms straight down, parallel to the torso. It is known that the human's position gravitates towards the neutral position.

3.7.2 Effort

Effort is measured as the cumulative displacement of the joints from their initial position, rather than from the neutral position. While discomfort is measured

Table 3.2 Joint Weights for Joint Displacement

Joint Variables	Joint Weight
q_1, q_4, q_7, q_{10}	100
q_2, q_5, q_8, q_{11}	100 when $q_i - q_i^N > 0$ 1000 when $q_i - q_i^N < 0$
q_3, q_6, q_9, q_{12}	5
q_{13}	75
q_{14}, q_{15}, q_{16}	1
q_{17}	50 when $q_i - q_i^N > 0$ 1 when $q_i - q_i^N < 0$
$q_{18}, q_{19}, q_{20}, q_{21}$	1

relative to a position that is deemed most comfortable, and the posture with the minimum discomfort gravitates towards the neutral position, effort is measured relative to a starting position, regardless of the starting position's comfort level. Consequently, effort depends greatly on the initial configuration of the limbs prior to motion and is most significant when a series of target points are selected with the posture changing from point to point. For an initial set of joint variables $q_i^{initial}$, a simple measure of effort is expressed as follows:

$$f_{Effort}(\mathbf{q}) = \sum_{i=1}^{n} \gamma_i (q_i - q_i^{initial})^2 \qquad (3.9)$$

where γ_i is a weight assigned to each joint and is comparable to w_i.

3.7.3 Delta potential energy

Potential energy is a well-understood basic concept and has been used successfully as an objective with robotic movements. However, implementing an energy function as a human performance measure requires special considerations. This proposed performance measure stems from difficulties with the above-mentioned joint displacement function and from deficiencies in an existing performance measure that depends on the potential energy of an arm (Abdel-Malek et al., 2001a–d; Mi et al., 2002b; Mi, 2004).

With joint displacement, the weights are set based on intuition and experimentation, and although the postures obtained by minimizing joint displacement are acceptable, the question arises as to whether or not there are more practical, less ad hoc approaches to setting the weights. The idea of potential energy provides one such alternative. With potential energy, the weights are essentially based on the mass of different segments of the body, and in a sense, an individual objective function is developed for each segment.

Whereas the previous potential-energy function incorporates only the potential energy of an arm, we consider the complete upper body. We represent the primary segments of the upper body with six lumped masses: three for the lower, middle, and upper torso, respectively; one for the upper arm; one for the forearm; and one for the hand. We then determine the potential energy for each mass. The actual masses for the segments are determined based on data from Chaffin and Anderson (1991). The heights of the masses, rather than the joint displacements, provide the components of the human performance measure. Mathematically, the weight (force of gravity) of a segment of the upper body provides a multiplier for movement of the segment in the vertical direction. The height of each segment is a function of the joint angles, so, in a sense, the weights of the lumped masses replace the scalar multipliers w_i, which are used in the joint displacement function.

If potential energy is used directly, as is the case with the previous potential-energy function, there is always a tendency for the avatar to simply bend over, thus reducing potential energy. All of the lumped masses gravitate towards the

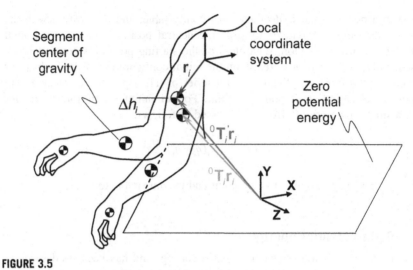

FIGURE 3.5

Potential energy.

same vertical baseline where the potential is considered zero. Consequently, we introduce the idea of minimizing the *change* in potential energy. It is calculated as follows. Each segment in the human model has a specified center of mass as depicted in Figure 3.5.

The vector from the origin of a link's local coordinate system to its center of mass is given by $\mathbf{r}_i$, where the subscript indicates the relevant local coordinate system. In order to determine the position of any part of the body, we use the DH transformation matrices $^{(i-1)}\mathbf{T}_i$. Note that $\mathbf{r}_i$ is actually an augmented 4×1 vector with respect to local coordinate system-i, rather than a 3×1 vector typically used with Cartesian space, as discussed in subsection 10.2.1; $\mathbf{g} = [0 \quad -g \quad 0 \quad 0]^T$ is the augmented gravity vector. When the avatar moves from one configuration to another, P_i' represents the potential energy of the initial configuration, and P_i represents the potential energy of the current configuration. The potential energy terms for the ith body part are $P_i' = m_i \mathbf{g}^T {}^0\mathbf{T}'_1 \cdots {}^{i-1}\mathbf{T}'_i \mathbf{r}_i$ and $P_i = m_i \mathbf{g}^T {}^0\mathbf{T}_1 \cdots {}^{i-1}\mathbf{T}_i \mathbf{r}_i$. In Figure 3.5, Δh_i is the y-component of the vector $^0\mathbf{T}'_1 \cdots {}^{i-1}\mathbf{T}'_i \mathbf{r}_i - {}^0\mathbf{T}_1 \cdots {}^{i-1}\mathbf{T}_i \mathbf{r}_i$. The final performance measure, which is minimized, is defined as follows:

$$f_{Delta-potential-energy}(\mathbf{q}) = \sum_{i=1}^{\kappa} (P_i - P_i')^2 \qquad (3.10)$$

Note that Equation (3.10) can be written in the form of a weighted sum as follows:

$$f_{Delta-potential-energy}(\mathbf{q}) = \sum_{i=1}^{\kappa} (m_i g)^2 (\Delta h_i)^2 \qquad (3.11)$$

where $(m_i g)^2$ represents the weights and $(\Delta h_i)^2$ acts as the individual objective functions; κ is the number of lumped masses. Potential energy is a relative quantity, and the term Δh_i usually refers to the vertical distance between a mass and a plane of zero potential, where the plane is the same for all masses. However, in this case, Δh_i is measured relative to the neutral position for a particular mass. Essentially, each mass has a different plane of zero potential. In this case, the initial position is the neutral position described in relation to joint displacement. Thus, with this performance measure, the human model again gravitates towards the neutral position, as was the case with joint displacement. However, horizontal motion of the lumped masses has no effect.

3.7.4 Discomfort

The idea of modeling discomfort can be somewhat ambiguous, as it is a subjective quantity, the evaluation of which may vary from person to person. However, it is possible to incorporate distinct factors that contribute to discomfort, though the actual absolute value for discomfort may not be significant. In this section, we present a human performance measure for musculoskeletal discomfort that incorporates three such factors:

1. The tendency to move different segments of the body sequentially
2. The tendency to gravitate to a reasonably comfortable neutral position
3. The discomfort associated with moving while joints are near their respective limits.

Some experimental work has been completed on discomfort in an effort to determine which factors contribute to discomfort, and the above-mentioned factors surface in these studies. However, these factors have not yet been collectively incorporated into an effective optimization-based human performance measure for use with virtual humans.

In order to incorporate the first factor (the tendency to move different segments of the body sequentially), there are several strategies to induce motion in a certain order, or with higher weighted joints than others.

For example, in an effort to reach a particular target point, one first uses one's arm. Then, only if necessary, does one bend the torso. If the target is still out of reach, one may extend the clavicle joint or take a step to reach it. The weights used to approximate the lexicographic approach are shown in Table 3.3 (Marler et al., 2005a,b, 2009).

The weights in Table 3.3 are used in a function that is based on Equation (3.7) with the neutral position defined as shown in Equation (3.8). In this way, the second factor of discomfort (the tendency to gravitate to a reasonably comfortable neutral position) is incorporated. Prior to applying the weights, each term in Equation (3.7) is normalized as follows:

$$\Delta q^{norm} = \frac{q_i - q_i^N}{q_i^U - q_i^L} \tag{3.12}$$

Table 3.3 Joint Weights for Discomfort	
Joint Variables	**Joint Weight**
$q_1, \ldots, q_{12}$	1×10^4
q_{13}, q_{14}	1×10^8
$q_{15}, \ldots, q_{21}$	1

With this normalization scheme, each term $(\Delta q_i^{norm})^2$ acts as an individual objective function and has values between zero and one.

Generally, this approach works well, but it often results in postures with joints extended to their limits, and such postures can be uncomfortable. To rectify this problem and to incorporate the final factor of discomfort (the discomfort associated with moving while joints are near their respective limits), specially designed penalty terms are added to the discomfort function such that discomfort increases significantly as joint values approach their limits. The final discomfort function is given as follows:

$$f_{Discomfort}(\mathbf{q}) = \frac{1}{G} \sum_{i=1}^{DOF} [\gamma_i (\Delta q_i^{norm}) + G \times QU_i + G \times QL_i] \qquad (3.13)$$

$$QU_i = \left(0.5 \, Sin \left(\frac{5.0(q_i^U - q_i)}{q_i^U - q_i^L} + 1.571 \right) + 1 \right)^{100} \qquad (3.14)$$

$$QL_i = \left(0.5 \, Sin \left(\frac{5.0(q_i - q_i^L)}{q_i^U - q_i^L} + 1.571 \right) + 1 \right)^{100} \qquad (3.15)$$

where $G \times QU$ is a penalty term associated with joint values that approach their upper limits, and $G \times QL$ is a penalty term associated with joint values that approach their lower limits; γ_i represents the weights defined in Table 3.3. Each term varies between zero and G, as $(q_i^U - q_i)/(q_i^U - q_i^L)$ and $(q_i - q_i^L)/(q_i^U - q_i^L)$ vary between zero and one. Figure 3.6 illustrates the curve for the following function, which represents the basic structure of the penalty terms

$$Q = (0.5 \, Sin(5.0r + 1.571) + 1)^{100} \qquad (3.16)$$

r represents either $(q_i^U - q_i)/(q_i^U - q_i^L)$ or $(q_i - q_i^L)/(q_i^U - q_i^L)$.

Thus, as Figure 3.6 illustrates, the penalty term has a value of zero until the joint value reaches the upper or lower 10% of its range, where either $(q_i^U - q_i)/(q_i^U - q_i^L) \leq 0.1$ or $(q_i - q_i^L)/(q_i^U - q_i^L) \leq 0.1$. The curve for the penalty term is differentiable and reaches its maximum of $G = 1 \times 10^6$ when $r = 0$. The final function in Equation (3.13) is multiplied by $1/G$ for the sake of presentation

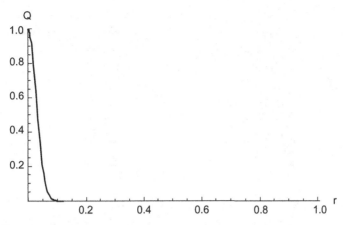

FIGURE 3.6

Graph of discomfort joint-limit penalty term.

Table 3.4 Objective Function Values with the Front Right Target

	Joint Displacement	Delta Potential Energy
Min. Joint Displacement	1.6	27
Min. Delta Potential Energy	176.	3

Table 3.5 Objective Function Values with the Back Target

	Joint Displacement	Delta Potential Energy
Min. Joint Displacement	2.2	32
Min. Delta Potential Energy	131	1.7

to the user, so discomfort does not have extremely high values when compared with other performance measures.

3.7.5 Single-objective optimization

Results of single cost function optimization are shown below using the front right and the back target. The values of the objective functions are shown in Tables 3.4 and 3.5.

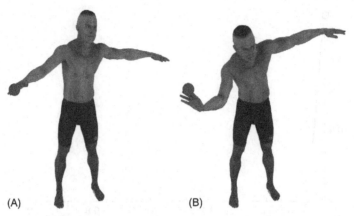

(A)   (B)

FIGURE 3.7 Front right target.

A. Posture calculated after minimizing joint displacement; B. Posture calculated after minimizing delta potential energy.

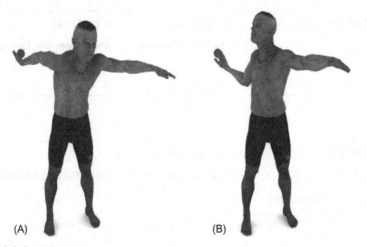

(A)   (B)

FIGURE 3.8 Back target.

A. Posture calculated after minimizing joint displacement; B. Posture calculated after minimizing delta potential energy.

Indeed, the two objective functions oppose each other: what decreases the value of one, increases the value of the other. In some cases, one function can dominate the aggregated objective function regardless of the weighting values.

The actual postures associated with the points in Tables 3.4 and 3.5 are shown in Figures 3.7 and 3.8. In evaluating the visual results, we are concerned primarily with gross movement; the nuances of skin deflection are not addressed. Generally, minimizing delta potential energy results in increased torso rotation about the vertical axis. This is because rotation about the y-axis does not alter the potential energy of any part of the body. Using delta potential energy also results in more bending of the elbow. As one can see from the results with the front right target, delta potential energy should not necessarily be used in place of joint displacement; it should not be used alone. This suggests that *human posture is not governed primarily by potential energy*, but as we show, potential energy does play a role. This finding is counterintuitive and consequently significant.

3.7.6 **Numerical solutions to optimization problems**

The subject of this book is the introduction of a new method called predictive dynamics. The method uses numerical optimization at its core. Optimization is a well-developed field, and many numerical methods and strategies have been researched to obtain solutions since the 1960s. The intent here is not to describe the details of numerical methods for solving optimization problems, but only to give a brief introduction to how such problems can be solved.

Based on extensive research on numerical optimization methods, a few promising methods for solution of practical problems have emerged: the sequential quadratic programming (SQP) method, the interior point (IP) methods, the exterior penalty methods, the augmented Lagrangian methods, and the generalized reduced gradient (GRG) methods. Some of these methods have been implemented into widely used software environments, such as MATLAB and Excel. However, these programs are cumbersome to use for digital human modeling applications. Several other commercial codes are available, and some of these are listed below. This is by no means a comprehensive list.

1. SNOPT: SNOPT stands for Sparse Nonlinear OPTimizer. It implements a sparse SQP algorithm that is suitable for large sparse optimization problems (Gill et al., 2002). It also has an algorithm to solve dense optimization problems.
2. KNITRO: KNITRO has three algorithms for solving nonlinear optimization problems: interior-point direct, interior-point conjugate gradient, and active set algorithms (Byrd et al., 2006).
3. CONOPT: CONOPT implements a large-scale GRG algorithm (Drud, 1992).
4. LANCELOT: LANCELOT implements an augmented Lagrangian method (Conn et al., 1992).

Real-time optimization requires more sophisticated optimization formulation and implementation. While we have been able to implement a real-time posture prediction method into Santos®, including physics makes the problem difficult. This topic will be briefly discussed in later chapters.

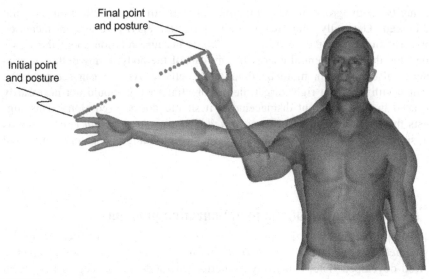

FIGURE 3.9

Posture predicted using initial and final points on the Cartesian path.

3.8 Motion between two points

Motion prediction is defined as calculating the joint variables versus time for each joint during motion of a kinematic skeleton from one point to another. The curve representing joint variables versus time is also called joint profiles.

In order to calculate the joint profiles in motion prediction, the initial and final value for each joint q_i must be known. Given a Cartesian path for the end-effector, these values are determined by running posture prediction for both the initial and final points along the path (Figure 3.9).

3.9 Joint profiles as B-spline curves

Joint profiles are defined as the curve of joint angle versus time. Indeed, each joint can be made of one or more DOF. Therefore, while we use the term *joint profiles*, each profile is indeed for each independent DOF.

Motion prediction involves finding joint profiles for each DOF in a kinematic model. While a motion is never discreet, it is important to have discrete points representing the joint profile. Therefore, we first introduce a "B-spline curve" as a linear combination of B-spline basis functions. B-spline curves are an ideal choice because they can be chosen to allow continuity, differentiability, and endpoint interpolation. Furthermore, B-spline curves give local control; hence, changing one coefficient affects only a small portion of the curve. This is desirable when using these coefficients as design variables in an optimization in order to shape the joint profile.

The pth-degree B-spline curve $C(u)$ is a linear combination of pth-degree basis functions $N_{k,p}(u)$, defined as follows:

$$C(u) = \sum_{k=0}^{n} N_{k,p}(u)P_k \qquad (3.17)$$

where

$$N_{k,p}(u) = \frac{u - u_k}{u_{k+p} - u_k} N_{k,p-1}(u) + \frac{u_{k+p+1} - u}{u_{k+p+1} - u_{k+1}} N_{k+1,\,p-1}(u), \qquad (3.18)$$

and

$$N_{k,0}(u) = \begin{cases} 1 & \text{if } u_k \leq u \leq u_{k+1} \\ 0 & \text{otherwise} \end{cases} \qquad (3.19)$$

Here, the curve has $n + 1$ coefficients P_k ($0 \leq k \leq n$). The vector $\mathbf{U} = \{u_0, \ldots, u_m\}$ is a non-decreasing sequence of real numbers called the knot vector; each u_i is a knot. The number of knots $m + 1$, the number of coefficients $n + 1$, and the degree p are related by

$$m = n + p + 1 \qquad (3.20)$$

For calculation of joint acceleration, the joint profiles must be at least twice differentiable; thus, it is necessary to use at least a third-degree B-spline curve. Furthermore, using a knot vector that begins and ends with multiplicity $p + 1$ will ensure that the joint trajectories interpolate the initial and final joint values. Hence, the joint profiles are defined by

$$q_i(u) = \sum_{k=0}^{n} N_{k,3}(u)P_k^{(i)} \qquad (3.21)$$

where q_i is the ith joint profile, $\{P_0^{(i)}, P_1^{(i)}, \ldots, P_n^{(i)}\}$ is the unknown coefficient vector for q_i, and $1 \leq i \leq n\text{DOF}$ (where $n\text{DOF}$ is the number of DOF). Hence, there are $(n + 1)*(n\text{DOF})$ coefficients to determine.

For example, for a 15-DOF kinematic model, the first and last coefficients are equal to the initial and final values; therefore, there are only $13 - 1 - 1 = 11$ coefficients per curve. Thus, there will be $(11)*(n\text{DOF}) = 11*15 = 165$ design variables for optimization.

Lastly, note that using B-spline curves gives a result with normalized time between zero and one second. The resulting motion can be scaled to longer durations.

3.10 **Motion prediction formulation**

The optimum motion along a path, given by a set of joint trajectories, is determined by solving the following optimization problem:

Find: $\mathbf{q}(u) = \{q_i(u) | 1 \leq i \leq n\mathrm{DOF}\}$
to minimize: $Discomfort$, etc.
subject to: $||\mathbf{x}^{end-eff}(\mathbf{q}(t_j)) - \mathbf{x}^{path}(t_j)||^2 \leq \varepsilon$(distance to path)   (3.22)
and $q_i^{Lower} \leq P_k^{(i)} \leq q_i^{Upper}$(joint limits)

where $\mathbf{q}(u)$ represents motion profiles for each joint. For this problem in Equation (3.22), the *feasible space* is defined as the set of all solutions $\mathbf{q}(u)$ for which every constraint is satisfied.

3.10.1 **Design variables**

As previously discussed, the design variables are the $(n + 1)*(n\mathrm{DOF})$ coefficients, $\{P_0^{(i)}, P_1^{(i)}, \ldots, P_n^{(i)}\}$ for $1 \leq i \leq n\mathrm{DOF}$. These coefficients shape the B-spline curve into optimal joint trajectories (which we call joint profiles). The resulting curve plots joint values (radians) versus time (seconds). The optimization requires an "initial guess" for the design variables. In this case, the coefficients for the ith joint are evenly spaced between the initial joint value q_i^{init} and the final joint value q_i^{final}. Hence, the initial guess for $q_i(u)$ becomes:

$$P_k^{(i)} = q_i^{init} + \frac{k}{n}(q_i^{final} - q_i^{init})$$ (3.23)

where there are $n + 1$ coefficients and $0 \leq k \leq n$. This guess satisfies the joint limits as long as the initial and final joint values also satisfy joint limits. However, it does not necessarily satisfy the distance constraint.

3.10.2 **Constraints**

The first constraint in Equation (3.22) is the distance constraint, which requires that the end-effector remain in contact with the given Cartesian path. The remaining constraints ascertain that each curve $q_i(u)$ lies between the upper and lower limits for that joint. Using the property of B-spline curves, restricting every coefficient $P_k^{(i)}$ to be within the joint limits will guarantee that the entire curve $q_i(u)$ lies within the joint limits. Enforcing these constraints helps ensure that the algorithm does not result in an unrealistic motion, and as a result will stay on the given path.

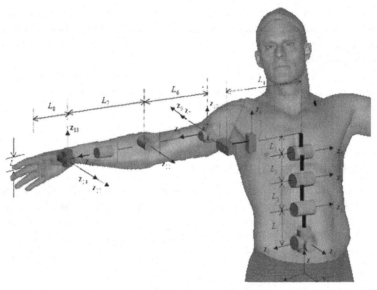

FIGURE 3.10

Detailed 15-DOF kinematic model of the torso and arm.

3.11 **A 15-DOF motion prediction**

The current 15-DOF model provides a simple example of optimization-based motion prediction. Although a higher-DOF model has the potential to improve results, it is useful to discuss the current model as a starting point and basis for comparison.

3.11.1 **The 15-DOF Denavit–Hartenberg model**

A 15-DOF kinematic chain for the upper torso and right arm was developed using the DH method (Figure 3.10). This model combines rotational and translational DOFs to estimate human joint movements. The distances L_i between joints define the transformation matrices of Equation (2.49). These values are considered input to the motion prediction routine. They can therefore be varied to correspond with specific anthropometric data. For the purpose of testing, reasonable values of L_i are determined by inspection.

Since the discomfort objective depends on a neutral posture, this 15-DOF model must have a defined neutral posture $\mathbf{q}^N = \{q_i^N | 1 \leq i \leq n\text{DOF}\}$. Currently, the chosen neutral posture is a standing position with the arms at the sides. The vector $\mathbf{q}^N$ is then defined as $q_{10}^N = 90.0$, $q_{13}^N = -90.0$, and $q_i^N = 0$ for all other i, $1 \leq i \leq n\text{DOF}$. Note that any q_i is measured with respect to the posture in Figure 3.10 and corresponds to the z_{i-1} axis. Furthermore, the positions of the target and end-effector must be given in terms of a common coordinate system. For this purpose, a global coordinate system is chosen to correspond to the *zeroth* coordinate frame of the model. Hence, the global position $(0,0,0)$ is located at the center of the torso. The end-effector for this model is a point on the thumb given by the local position $(L_8, -L_9, 0)$ relative to the last coordinate frame.

3.12 Optimization algorithm

To solve this problem we have used a number of commercial optimization software programs. However, these programs have consistently resulted in a violated distance constraint. In an effort to compensate for this, the distance constraint is combined with the objective and given a large weighting factor. This forms the following unconstrained objective:

$$f(\mathbf{q}) = f_{discomfort} + 1000.0^* \sum_{j=1}^{nTIME} \left\| \mathbf{x}^{end-eff}(\mathbf{q}(t_j)) - \mathbf{x}^{path}(t_j) \right\|^2 \qquad (3.24)$$

Furthermore, additional terms to control the shape of the resulting joint profile are added to the objective function. The first is a term to control the inconsistency of the curve—that is, to prevent joint curves that produce back and forth movements. This inconsistency is mathematically defined by:

$$f_{inconsistancy}(\dot{\mathbf{q}}) = \sum_{j=1}^{nTIME} \left(\sum_{i=1}^{nDOF} (|sign(\dot{q}_i(t_j)) - trend_i| + 1)|\dot{q}_i(t_j)| \right) \qquad (3.25)$$

where

$$sign(\dot{q}_i(t)) = \begin{cases} 1 & \text{if } \dot{q}_i(t) \geq 0 \\ -1 & \text{if } \dot{q}_i(t) < 0 \end{cases} \qquad (3.26)$$

and

$$trend_i = \begin{cases} 1 & \text{if } (q_i^f - q_i^0) \geq 0 \\ -1 & \text{if } (q_i^f - q_i^0) < 0 \end{cases} \qquad (3.27)$$

The second is a term to ensure smooth joint movement by minimizing the second derivative of the joint profile. The non-smoothness added to the objective is given by:

$$f_{nonsmoothness}(\ddot{q}) = \sum_{j=1}^{nTIME} \left(\sum_{i=1}^{nDOF} (\ddot{q}_i(t_j))^2 \right) \qquad (3.28)$$

Finally, the objective function for the unconstrained optimization problem becomes:

$$f(\mathbf{q}, \dot{\mathbf{q}}, \ddot{\mathbf{q}}) = w_1 f_{discomfort} + w_2 f_{inconsistnacy} + w_3 f_{nonsmoothness}$$
$$+ 1000.0 * \sum_{j=1}^{nTIME} ||\mathbf{x}^{end-eff}(\mathbf{q}(t_j)) - \mathbf{x}^{path}(t_j)||^2 \tag{3.29}$$

Mi (2004) describes the objective terms in Equations (3.25) and (3.28) in greater detail.

The limits for the generalized coordinates q_i are determined based on rough estimates and observation. However, these joint limits are input parameters to the optimization and can correspond to any anthropometric specifications. The values for the weights in Equation (3.29) are determined based on trial-and-error experiments.

3.13 Motion prediction of a 15-DOF model

Applying the current motion prediction algorithm to the 15-DOF model results in the motion shown in Figures 3.11 and 3.12. The motion was generated using an initial posture with zero rotation and an approximate target point of (41, −55, 32). Since the minimum jerk model is well accepted for point-to-point motion (Flash and Hogan, 1985), the end-effector follows a minimum jerk Cartesian path for this example. The optimization used 43 discretized Cartesian points along the path to constrain the end-effector. The five screenshots show the instantaneous postures at $t = 0.0$, 0.25, 0.5, 0.75, and 1.0 seconds. The calculation took about 17 seconds on a 1.8-GHz Pentium4 CPU with 512-MB RAM.

Figure 3.13 provides the some of the joint trajectories from the motion prediction. By inspecting the curve, it is apparent that the joint movements are smooth and continuous; they are devoid of abrupt changes in velocity or acceleration. However, note that the initial and final velocities are not necessarily zero. For an isolated point-to-point movement, restricting the initial and final velocities is intuitively desirable.

Furthermore, this motion is slightly unnatural, with excessive torso movement. As discussed earlier, it was necessary to combine the distance constraint with the

FIGURE 3.11

Motion prediction on 15-DOF model at five time instants.

FIGURE 3.12

Motion prediction on 15-DOF model at five instants, overlapped.

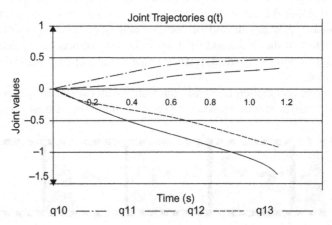

FIGURE 3.13

Joint trajectories (motion profiles) predicted by motion prediction on a 15-DOF model.

objective function because the DOT does not handle the constrained optimization well. In fact, the optimization yields similar results with this target point when the objective terms are omitted entirely. This suggests that the distance term in Equation (3.24) dominates the objective. However, if the multiplier (1000) is reduced to rectify this problem, then the constraint is not adequately satisfied. Conceptually and practically, the objective needs to be treated independently. Furthermore, using a larger-DOF model that allows more realistic movement in the upper torso and clavicle will improve the results.

3.14 Multi-objective problem statement

In this section, we formulate the posture prediction optimization problem using the above-described model. MOO is used first to develop new human performance measures. It is then used to combine these measures that serve as multiple objective functions in the final optimization formulation.

3.15 Design variables and constraints

As suggested earlier, the design variables for the final MOO problem are q_i, which indicate joint angles in units of degrees. The vector $\mathbf{q}$ represents the consequent posture. Because listing values for all of the joint angles with each predicted posture can be cumbersome and unrevealing, and therefore, results in the design space are depicted with actual pictures of the avatar.

The first constraint, called the distance constraint, requires the end-effector to contact the target point. There is one distance constraint for each end-effector. In addition, each generalized coordinate is constrained to lie within predetermined limits; q_i^U represents the upper limit for q_i, and q_i^L represents the lower limit. These limits ensure that the virtual human does not assume an unrealistic posture given the nature of actual human joints. Finally, we will demonstrate that the orientation of a body part can be controlled by implementing an additional *orientation constraint*.

3.16 Concluding remarks

This chapter has introduced the concept of posture prediction as an effective and robust method that yields natural human motion.

The main concept from this chapter is the use of optimization to solve human motion problems. We have shown how optimization lends itself well to answering the questions: "What would a human do under these constraints?" and "What is the best solution given those constraints?"

Posture prediction using optimization-based methods is effective in addressing the prediction of human motion and yields the following results:

a. Predicts natural human motion; this is due to the use of human performance measures as cost functions in an optimization algorithm.
b. Models with high fidelity; the DH representation method allows high-fidelity modeling that employs a high-DOF representation of the human.
c. Is suitable for computational implementation; the method and accompanying algorithm are suitable for computational methods yielding highly efficient algorithms.
d. Induces behavior; one of the most important aspects of this method is its ability to change the combination of cost functions to induce various behaviors.
e. Runs in real time and is therefore interactive; the method is computationally efficient and provides immediate interactivity for a user to posture the digital avatar.

References

Abdel-Malek, K., Yu, W., Jaber, M., 2001a. Realistic Posture Prediction. 2001 SAE Digital Human Modeling and Simulation.

Abdel-Malek, K., Wei, Y., Mi, Z., Tanbour, E., Jaber, M., 2001b. Posture prediction versus inverse kinematics. In Proceedings of the 2001 ASME Design Engineering Technical Conferences and Computers and Information in Engineering Conference (pp. 37−45). Pittsburgh, PA.

Abdel-Malek, K., Yang, J., Brand, R., Tanbour, E., 2001c. Towards understanding the workspace of the upper extremities. SAE Trans. J. Passenger Cars: Mech. Syst. 110.6, 2198−2206.

Abdel-Malek, K., Yu, W., Jaber, M., Duncan, J., (2001d). Realistic posture prediction for maximum dexterity. SAE Technical Paper 2001-01-2110. doi: 10.427/2001-01-2110.

Abdel-Malek, K., Yang, J., Brand, R., Tanbour, E., 2004a. Towards understanding the workspace of human limbs. Ergonomics 47 (13), 1386−1406.

Abdel-Malek, K., Yang, J., Yu, W., Duncan, J., 2004b. Human performance measures: mathematics. Proceedings of the ASME Design Engineering Technical Conferences (DAC 2004), Salt Lake City, UT.

Abdel-Malek, K., Yang, J., Mi, Z., Patel, V. C., Nebel, K., 2004c. Human upper body motion prediction. Proceedings of Conference on Applied Simulation and Modeling (ASM) 2004, Rhodes, Greece, pp. 28-30.

Abdel-Malek, K., Yang, J., Marler, T., Beck, S., Mathai, A., Zhou, X., et al., 2006. Towards a new generation of virtual humans Int. J. Human Fact. Model. Simul. 1(1), 2–39.

Abdel-Malek, K., Mi, Z., Yang, J., Nebel, K., 2005. Optimization-based layout design. J. Appl. Bionics Biomech. 2 (3/4), 187−196.

Arora, J.S., 2012. Introduction to Optimum Design, third ed. Elsevier Academic Press, San Diego, CA.

Byrd, R.H., Nocedal, J., Waltz, R.A., 2006. KNITRO: an integrated package for nonlinear optimization. In: di Pillo, G., Roma, M. (Eds.), Large-Scale Optimization. Springer-Verlag, pp. 35−59.

Case, K., Porter, J.M., Booney, M.C., 1990. SAMMIE: a man and workplace modelling system. In: Karwowski, W., Genaidy, A.M., Asfour, S.S. (Eds.), Computer-Aided Ergonomics. Taylor and Francis, New York, pp. 31–56.

Chaffin, D.B., Anderson, D.B.J., 1991. Occupational Biomechanics. Wiley, New York, NY.

Conn, A.R., Gould, N.I.M., Toint, P.L., 1992. LANCELOT—a fortran package for large-scale nonlinear optimization. Springer Series in Computational Mathematics. Springer-Verlag.

Drud, A.S., 1992. CONOPT—a large-scale GRG code. ORSA J. Comput. 6, 207–216.

Flash, T., Hogan, N., 1985. The coordination of arm movements: an experimentally confirmed mathematical model. J. Neurosci. 5(7), 1688–1703.

Gill, P.E., Murray, W., Saunders, M.A., 2002. SNOPT: an SQP algorithm for large-scale constrained optimization. Siam J Optim. 12 (4), 979–1006.

Jung, E.S., Choe, J., 1996. Human reach posture prediction based on psychophysical discomfort. Int. J. Ind. Ergon. 18, 173–179.

Marler, R.T., Rahmatalla, S., Shanahan, M., Abdel-Malek, K., 2005a. A new discomfort function for optimization-based posture prediction. Paper presented at the SAE Human Modeling for Design and Engineering Conference, Iowa City, IA.

Marler, R.T., Yang, J., Arora, J.S., Abdel-Malek, K., 2005b. Study of bi-criterion upper body posture prediction using pareto optimal sets. Paper presented at the IASTED International Conference on Modeling, Simulation, and Optimization, Oranjestad, Aruba.

Marler, R.T., Yang, J., Rahmatalla, S., Abdel-Malek, K., Harrison, C., 2009. Use of multi-objective optimization for digital human posture prediction. Eng. Optim. 41 (10), 925–943.

Mi, Z., 2004. Task-Based Prediction of Upper Body Motion, Ph.D. Dissertation, University of Iowa, Iowa City, IA.

Mi, Z., Yang, J., Abdel-Malek, K., Jay, L., 2002a. Planning for kinematically smooth manipulator trajectories. In Proceedings of the 2002 ASME Design Engineering Technical Conferences and Computer and Information in Engineering Conference (pp. 1065–73). New York: American Society of Mechanical Engineers.

Mi, Z., Yang, J., Abdel-Malek, K., Mun, J.H., Nebel, K., 2002b. Real-Time Inverse Kinematics for Humans. Proceedings of the 2002 ASME Design Engineering Technical Conferences and Computer and Information in Engineering Conference, 5A, September, Montreal, Canada, American Society of Mechanical Engineers, New York, 349–59.

Mi, Z., Jingzhou, Y., Abdel-Malek, K., 2009. Optimization-based posture prediction for human upper body. Robotica 27 (4), 607.

Porter, J.M., Case, K., Bonney, M.C., 1990. Computer workspace modeling. In: Wilson, J. R., Corlett, E.N. (Eds.), Evaluation of Human Work. Taylor and Francis, London, UK, pp. 472–499.

Yang, J., Abdel-Malek, K., Nebel, K., 2004a. Restrained and unrestrained driver reach barriers. SAE Trans. J. Aerosp. 113 (1), 288–296.

Yang, J., Abdel-Malek, K., Farrell, K., Nebel, K., 2004b. The IOWA interactive digital-human virtual environment. Paper presented at the 3rd Symposium on Virtual Manufacturing and Application, Anaheim, CA.

Yang, J., Marler, R.T., Kim, H., Arora, J.S., Abdel-Malek, K., 2004c. Multi-objective optimization for upper body posture prediction. Paper presented at the 10th AIAA/ISSMO Multidisciplinary Analysis and Optimization Conference, Albany, NY.

Yang, J., Marler, R., Kim, H., Arora, J., Abdel-Malek, K., 2004d. Multi-objective Optimization for Upper Body Posture Prediction. 10th AIAA/ISSMO Multidisciplinary Analysis and Optimization Conference, August, Albany, NY, American Institute of Aeronautics and Astronautics, Washington, DC.

Yang, J., Pena-Pitarch, E., Kim, J., Abdel-Malek, K., 2006a. Posture prediction and force/torque analysis for human hand (SAE Paper No. 2006-01-2326). In Proceedings of the SAE 2006 Digital Human Modeling for Design and Engineering Conference.

Yang, J., Sinokrot, T., Abdel-Malek, K., Nebel, K., 2006b. Optimization-based workspace zone differentiation and visualization for Santos (SAE Paper No. 2006-01-0696). In Proceedings of the SAE 2006 Digital Human Modeling for Design and Engineering Conference.

Yang, J., Marler, T., Beck, S., Abdel-Malek, K., Kim, H. -J., 2006c. Real-time optimal-reach posture prediction in a new interactive virtual environment. J. Comput. Sci. Technol. 21 (2), 189−198.

Yang, J., Marler, T., Beck, S., Kim, J., Wand, Q., Zhou, X., et al., 2006d. New capabilities for the virtual human Santos. Paper presented at the SAE 2006 World Congress, Detroit, MI.

Yang, Q., Han, R.P.S., Frey Law, L.A., 2006. Simulating motor units for fatigue in arm muscles in digital humans. Paper presented at the 2006 Digital Human Modeling for Design and Engineering Conference, SAE, Lyon, France.

Yu, W., 2001. Optimal Placement of Serial Manipulators, Ph.D. Dissertation, University of Iowa, Iowa City, IA.

Zacher, I., Bubb, H., 2004. Strength based discomfort model of posture and movement. SAE Digital Human Modelling for Design and Engineering, Rochester, MI. SAE paper 2004-01-2139.

Recursive Dynamics

Never mistake motion for action.
Ernest Hemingway (1899–1961)

4.1 Introduction

The aim of this chapter is to introduce and develop a rigorous computational platform for describing the dynamics of a system of segmented links of a human body. Mass, moments of inertia, velocities, and accelerations will form a coupled set of nonlinear differential equations called the equations of motion.

General dynamics equations have been extensively studied in recent decades in terms of computational efficiency as the advent of computational platforms has readily enabled the calculation of physics for objects in motion. A convenient method for formulating the equations of motion, building upon the Denavit–Hartneberg (DH) method and using the Lagrangian equations for a serial kinematic chain, will be presented. This methodology lends itself well for kinematic skeletons of the human body and yields a convenient and compact description of the equations of motion. However, computation of the torques from equations of motion is of the order $O(n^4)$, where n is the number of degrees of freedom (DOF) for the system.

The human motion prediction optimization problem is usually a large-scale sparse nonlinear programming (NLP) problem. Accurate sensitivity (gradient calculation) is a key factor to efficiently achieve an optimal solution. Although the finite difference approach can be used to approximate gradients, the computational expense becomes substantial as the number of variables increases (i.e., the number of DOF). In addition, accuracy of the derivatives can affect convergence of the optimization process, thus leading to further computational expense.

Although different algorithms for sensitivity of dynamic equations have been studied, limited work is found for inverse recursive Lagrangian formulation with sensitivity for general motion planning problems. By using 4×4 transformation matrices (DH method), recursive Lagrangian formulation is more efficient and convenient to implement compared with recursive Newton–Euler formulation (Hollerbach, 1980). In addition, the Lagrangian formulation is the energy concept for the equations of motion typically defined in the joint space; this results in more convenient calculation of joint torques, especially for a skeletal model, compared with the Newton–Euler formulation, which is based on the Cartesian coordinates. Recursive dynamics will be introduced and sensitivity analysis will be derived.

In this chapter, the recursive Lagrangian dynamics and sensitivity formulation for the system are presented. Implementation aspects of sensitivity calculations for the complex articulated human mechanism are addressed. The developed formulation can systematically treat open-loop, closed-loop and branched mechanical systems. In addition, the sensitivity analysis needed for the optimization process is easier to implement. The formulation is based on DH transformation matrices and external forces, and moments at any point of the mechanism are included in the recursive formulation.

In order to demonstrate this formulation, an optimal time trajectory planning problem for a two-link human arm model is solved. Initial and final conditions are given. Total travel time is minimized subject to the joint torque limits.

We begin by developing the equations of motion. We first formulate a general static torque equation in vector form. This chapter is an adaptation of our work as it has appeared in Xiang et al. (2009a,b, 2010a,b,c).

4.2 General static torque

To obtain a relationship for the torque in terms of linear and angular velocity vectors, we use the chain rule in joint space as follows. For the vector of angular joint variables $\mathbf{q} = [q_1 \ q_2 \ldots q_n]^T$; the Jacobian ($\mathbf{J(q)}$) provides a direct relationship between the velocity in joint and Cartesian spaces.

$$\begin{bmatrix} \mathbf{v} \\ \boldsymbol{\omega} \end{bmatrix} = \mathbf{J(q)}\dot{\mathbf{q}} \qquad (4.1)$$

where, $\mathbf{v}$ is the translational velocity of the end-effector and $\boldsymbol{\omega}$ is the angular velocity of the end-effector frame. $\mathbf{J(q)}$ is the augmented Jacobian matrix (Sciavicco and Siciliano, 2000) of the kinematic structure defined by

$$\mathbf{J(q)} = \begin{bmatrix} \mathbf{J}_x \\ \mathbf{J}_\omega \end{bmatrix} \qquad (4.2)$$

This also indicates that the virtual displacements have the similar relationship:

$$\begin{bmatrix} \delta\mathbf{x} \\ \delta\boldsymbol{\theta} \end{bmatrix} = \mathbf{J(q)}\delta\mathbf{q} \quad \forall \delta\mathbf{q} \qquad (4.3)$$

where $\delta\mathbf{x}$, $\delta\boldsymbol{\theta}$, and $\delta\mathbf{q}$ are the virtual displacement vectors of Cartesian linear, Cartesian angular, and joint variables, respectively.

We begin with the calculation of the torque at each joint. To account for all of the elements that enter into calculating the torque at a given joint, we apply the principle of virtual work to the two force systems. As for the joint torques, its associated virtual work is

$$\delta W_\tau = \tau^T \delta \mathbf{q} \tag{4.4}$$

where δW is the virtual work and τ^T denotes the transpose of τ. For the end-effector forces $\mathbf{F} = [\mathbf{f}^T \quad \mathbf{m}^T]^T$, comprised of a force vector $\mathbf{f}$ and a moment vector $\mathbf{m}$, the virtual work performed is

$$\delta W_\mathbf{F} = \mathbf{f}^T \delta \mathbf{x} + \mathbf{m}^T \omega \delta t \tag{4.5}$$

where $\delta \mathbf{x}$ is the linear virtual displacement and $\omega \delta t$ is the angular virtual displacement of the end-effector, respectively. Because the difference between the virtual work of the joint torques and the virtual work of the end-effector forces shall be null for all joint virtual displacements, we write

$$\tau^T \delta \mathbf{q} = \mathbf{F}^T \mathbf{J}(\mathbf{q}) \delta \mathbf{q} \quad \forall \mathbf{q} \tag{4.6}$$

The relationship between the joint torque vector and end-effector force/moment vector is then given by

$$\tau = \mathbf{J}^T \mathbf{F} \tag{4.7}$$

where the torque vector is $\tau = [\tau_1, \tau_2, \ldots, \tau_n]^T$.

Now, we extend this formulation to the case where multiple external loads (both translational and rotational) are applied to any location of any link, not necessarily to the end-effector. Let's assume that a general form of external load $\mathbf{F}_k$ is applied to the point at $^k r_k$ location of link k, where $^k r_k$ location vector is expressed with respect to k^{th} local coordinate frame.

This point of application of external load can be regarded as the end-effector for the corresponding external load. The augmented Jacobian matrix $\mathbf{J}_k$ for this point is derived from the linear relationship between the joint velocity vector and the Cartesian velocity vector:

$$\mathbf{J}_k(\mathbf{q}) = \begin{bmatrix} \dfrac{\partial^0 \mathbf{T}_1(\mathbf{q})}{\partial q_1} {}^k\mathbf{r}_k & \cdots & \dfrac{\partial^0 \mathbf{T}_i(\mathbf{q})}{\partial q_i} {}^k\mathbf{r}_k & \cdots & \dfrac{\partial^0 \mathbf{T}_k(\mathbf{q})}{\partial q_k} {}^k\mathbf{r}_k \\ \mathbf{Z}_0(\mathbf{q}) & \cdots & \mathbf{Z}_{i-1}(\mathbf{q}) & \cdots & \mathbf{Z}_{k-1}(\mathbf{q}) \end{bmatrix}_{6 \times k} \tag{4.8}$$

where, $i = 1, \ldots, k$ is the local z-axis vector of joint i expressed in terms of the global coordinate system.

Therefore the joint torque vector due to the external load applied at point $^k r_k$ of link k is

$$\tau_k = \mathbf{J}_k^T \mathbf{F}_k \tag{4.9}$$

From the principle of superposition, the total joint torques due to several external loads is obtained as a sum of all joint torques:

$$\tau = \sum_k \mathbf{J}_k^T \mathbf{F}_k \tag{4.10}$$

4.3 Dynamic equations of motion

For the purpose of this discussion, consider a kinematic skeleton of a human body represented as an articulated (jointed) set of rigid bodies as depicted in Figure 4.1. A global coordinate system is established at the base of the chain, and a local coordinate system at each link as mandated by the DH method.

The general form of dynamic equation of motion is derived from Lagrange's equation. Assuming that only the gravity forces and the driving joint torques are applied to the human links, the Lagrange's equation is written as follows:

$$\frac{d}{dt}\frac{\partial L}{\partial \dot{q}_i} - \frac{\partial L}{\partial q_i} = \tau_i, i = 1, \ldots, n \tag{4.11}$$

where $L = T - V$ is called the Lagrangian. T is the total kinetic energy, V is the total potential energy, q_i is the generalized coordinate of joint i, τ_i is the generalized torque of joint i, n is the number of the total DOF, and t is the time. Here, τ_i is the driving torque actuated by human muscles.

The velocity of the endpoint can be derived as follows:

$$\mathbf{v}_i = \frac{d}{dt}(^0\mathbf{r}_i) = \frac{d}{dt}(^0\mathbf{T}_i{}^i\mathbf{r}_i) = \left(\sum_{j=1}^{i} \frac{\partial {}^0\mathbf{T}_j(\mathbf{q})}{\partial q_j} \dot{q}_j\right) {}^i\mathbf{r}_i \tag{4.12}$$

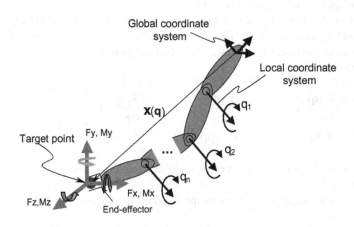

FIGURE 4.1

Joint-link system with external loads showing the generalized variables—the vector $\mathbf{x}(\mathbf{q})$ and end-effector—with respect to the global coordinate system.

The kinetic energy K_i of link i is calculated from the kinetic energy of a differential mass element as

$$K_i = \frac{1}{2} \int_m Tr(\mathbf{v}_i \mathbf{v}_i^T) dm \qquad (4.13)$$

where, Tr is the trace of a matrix, i.e., the sum of the diagonal elements of a matrix.

Therefore, the total kinetic energy K of the human-link system is

$$K = \sum_{i=1}^{n} K_i = \frac{1}{2} \sum_{i=1}^{n} \int_m Tr(\mathbf{v}_i \mathbf{v}_i^T) dm \qquad (4.14)$$

Expanding the velocities and expressing them in terms of joint velocities $\dot{q}_j$ using Equation (4.12), the total kinetic energy can be expressed in terms of joint space as

$$K = \frac{1}{2} \dot{\mathbf{q}}^T \mathbf{M}(\mathbf{q}) \dot{\mathbf{q}} = \frac{1}{2} \sum_{i=1}^{n} \sum_{k=1}^{n} M_{ik} \dot{q}_i \dot{q}_k \qquad (4.15)$$

where $\dot{\mathbf{q}}$ is the joint velocity vector and M_{ik} is the (i, k) element of the mass-inertia symmetric matrix $\mathbf{M}(\mathbf{q})$ such that

$$M_{ik}(\mathbf{q}) = \sum_{j=\max(i,k)}^{n} Tr\left(\frac{\partial^0 \mathbf{T}_j(\mathbf{q})}{\partial q_k} \mathbf{I}_j \left[\frac{\partial^0 \mathbf{T}_j(\mathbf{q})}{\partial q_i}\right]^T\right) \quad i,k = 1, 2, \ldots, n \qquad (4.16)$$

and $\mathbf{I}_j$ is the inertia matrix as below:

$$\mathbf{I}_i = \begin{bmatrix} \dfrac{-I_{xx} + I_{yy} + I_{zz}}{2} & -I_{xy} & -I_{xz} & m_i \bar{x}_i \\ -I_{xy} & \dfrac{I_{xx} - I_{yy} + I_{zz}}{2} & -I_{yz} & m_i \bar{y}_i \\ -I_{xz} & -I_{yz} & \dfrac{I_{xx} + I_{yy} - I_{zz}}{2} & m_i \bar{z}_i \\ m_i \bar{x}_i & m_i \bar{y}_i & m_i \bar{z}_i & m_i \end{bmatrix} \qquad (4.17)$$

where m_i is the mass of link i, $(\bar{x}_i, \bar{y}_i, \bar{z}_i)$ is the location vector of center of mass of link i, expressed in terms of ith coordinate frame, $I_{xx}, \ldots, I_{xy}, \ldots$ are the moments/products of inertia of link i with respect to i^{th} coordinate frame.

The potential energy P_i of each link i is

$$P_i = -m_i \mathbf{g}^T {}^0\bar{\mathbf{r}}_i = -m_i \mathbf{g}^T ({}^0\mathbf{T}_i {}^i\bar{\mathbf{r}}_i), \quad i = 1, 2, \ldots, n \qquad (4.18)$$

where ${}^i\bar{\mathbf{r}}_i$ is the center of mass vector of link i with respect to the ith local coordinate frame and $\mathbf{g}$ is the augmented 4×1 gravity vector.

Therefore, the total potential energy P due to gravity is the sum of each link's potential energy:

$$P = \sum_{i=1}^{n} P_i = \sum_{i=1}^{n} - m_i \mathbf{g}^T ({}^0\mathbf{T}_i{}^i\bar{\mathbf{r}}_i) \tag{4.19}$$

The kinetic energy Equation (4.15) and the potential energy Equation (4.19) are substituted into the Lagrange's equation (Equation 4.11) to derive the equations of motion, based on the procedure given in Fu et al. (1987).

As a result, the equation of motion is obtained as follows:

$$\tau = \mathbf{M}(\mathbf{q})\ddot{\mathbf{q}} + \mathbf{V}(\mathbf{q},\dot{\mathbf{q}}) + \sum_i \mathbf{J}_i{}^T m_i \mathbf{g} \tag{4.20}$$

where, $\mathbf{V}(\mathbf{q},\dot{\mathbf{q}})$ is the Coriolis and Centrifugal torque vector

$$\mathbf{V}_i(\mathbf{q},\dot{\mathbf{q}}) = \sum_{k=1}^{n} \sum_{m=1}^{n} \sum_{j=\max(i,k,m)}^{n} Tr \left(\frac{\partial^{2} {}^0\mathbf{T}_j(\mathbf{q})}{\partial q_k \partial q_m} \mathbf{I}_j \left[\frac{\partial \, {}^0\mathbf{T}_j(\mathbf{q})}{\partial q_i} \right]^T \right) \dot{q}_k \, \dot{q}_m, \quad i,k,m = 1,2,\ldots,n \tag{4.21}$$

and $\sum_i \mathbf{J}_i{}^T m_i \mathbf{g}$ is the joint torque vector due to gravity force.

Again, from the principle of superposition, the general equation of motion including several external loads can be obtained by adding the restoring torque term (Equation 4.20) and static torque term (Equation 4.10) to the above equation (Equation 4.20). As a result, the final version of general equation of motion with external loads, in vector-matrix form is

$$\tau = \underbrace{\mathbf{M}(\mathbf{q}) \, \ddot{\mathbf{q}}}_{massinertia\ matrix} + \underbrace{\mathbf{V}(\mathbf{q},\bar{\mathbf{q}})}_{CoriolisCentrifugal} + \underbrace{\sum_i \mathbf{J}_i{}^T m_i \mathbf{g}}_{gravity - forces} + \underbrace{\sum_k \mathbf{J}_k{}^T \mathbf{F}_k}_{external - forces} + \underbrace{\mathbf{K}(\mathbf{q} - \mathbf{q}^N)}_{muscle - elasticity} \tag{4.22}$$

Equation 4.22 is the most common form of the equations of motion as each terms is readily identifiable. Note that these equations govern the motion and are highly nonlinear.

4.4 Formulation of regular Lagrangian equation

The regular form of the Lagrangian equation can be written in vector-matrix form (Fu et al., 1987):

$$\tau = \mathbf{M}(\mathbf{q}) \, \ddot{\mathbf{q}} + \mathbf{V}(\mathbf{q}, \dot{\mathbf{q}}) + \sum_i \mathbf{J}_i^T m_i \mathbf{g} + \mathbf{J}_s^T \mathbf{f}_s \tag{4.23}$$

where

$$M_{ik}(\mathbf{q}) = \sum_{j=\max(i,k)}^{n} Tr\left(\frac{\partial\, {}^{0}\mathbf{T}_{j}(\mathbf{q})}{\partial q_{k}} \mathbf{I}_{j} \left(\frac{\partial\, {}^{0}\mathbf{T}_{j}(\mathbf{q})}{\partial q_{i}}\right)^{T}\right) \quad i,k = 1,2,\ldots,n \qquad (4.24)$$

$$V_{i}(\mathbf{q},\dot{\mathbf{q}}) = \sum_{k=1}^{n}\sum_{m=1}^{n}\sum_{j=\max(i,k,m)}^{n} Tr\left(\frac{\partial^{2}\, {}^{0}\mathbf{T}_{j}(\mathbf{q})}{\partial q_{k}\partial q_{m}} \mathbf{I}_{j} \left(\frac{\partial\, {}^{0}\mathbf{T}_{j}(\mathbf{q})}{\partial q_{i}}\right)^{T}\right) \dot{q}_{k}\dot{q}_{m} \quad i,k,m = 1,2,\ldots,n$$

$$(4.25)$$

$$\mathbf{J}_{i} = \frac{\partial\, {}^{0}\mathbf{T}_{i}}{\partial q_{i}}\bar{r}_{i} \qquad (4.26)$$

$$\mathbf{J}_{s} = \frac{\partial\, {}^{0}\mathbf{T}_{s}}{\partial q_{s}}\bar{r}_{s} \qquad (4.27)$$

where $\mathbf{J}_{i}$ is the Jacobian matrix for link i and $\bar{r}_{i}$ is the position of the center of mass of link i with respect to the ith coordinate system; $\mathbf{I}_{j}$ is the augmented inertia matrix for link j; $\mathbf{J}_{s}$ is the Jacobian matrix for the external load $\mathbf{f}_{s}$ and $\bar{r}_{s}$ is the position where external load is applied in the sth coordinate system.

4.4.1 Sensitivity analysis

Sensitivity analysis means calculation of derivatives of various quantities with respect to the state variables. We note from Equations (4.6−4.10) that the regular Lagrangian equations are coupled, nonlinear, and second-order differential equations. $\mathbf{M}(\mathbf{q})$ is an $n \times n$ matrix and $\mathbf{V}(\mathbf{q},\dot{\mathbf{q}})$ is an $n \times 1$ vector. Each term involves summation and state variables. Direct sensitivity analysis gives the $n \times n$ sensitivity matrix as

$$\frac{\partial \boldsymbol{\tau}}{\partial \mathbf{q}} = \frac{\partial \mathbf{M}(\mathbf{q})\,\ddot{\mathbf{q}}}{\partial \mathbf{q}} + \frac{\partial \mathbf{V}(\mathbf{q},\dot{\mathbf{q}})}{\partial \mathbf{q}} + \frac{\partial \sum_{i} \mathbf{J}_{i}^{T} m_{i}\mathbf{g}}{\partial \mathbf{q}} + \frac{\partial \mathbf{J}_{s}^{T}\mathbf{f}_{s}}{\partial \mathbf{q}} \qquad (4.28)$$

4.5 Recursive Lagrangian equations

We begin with a general formulation for forward recursive kinematics. Recursive dynamics methods have been shown to allow for efficient simulation of dynamic systems with large DOFs regardless of whether they are open, closed, or branched loops. Recursive dynamics averts the need for the reformulation of the dynamic equations for human systems. Furthermore, recursive dynamics provide for an increased stability in numerical performance.

4.5.1 Forward recursive kinematics

We can define 4×4 matrices $\mathbf{A}_j$, $\mathbf{B}_j$, $\mathbf{C}_j$ as recursive position, velocity, and acceleration transformation matrices, respectively, for the jth joint. Given the link transformation matrix $(\mathbf{T}_j)$ and the kinematics state variables for each joint (q_j, $\dot{q}_j$, and $\ddot{q}_j$), we have for $j = 1$ to n:

$$\mathbf{A}_j = \mathbf{T}_1 \mathbf{T}_2 \mathbf{T}_3 \cdots \mathbf{T}_j = \mathbf{A}_{j-1} \mathbf{T}_j \tag{4.29}$$

$$\mathbf{B}_j = \dot{\mathbf{A}}_j = \mathbf{B}_{j-1} \mathbf{T}_j + \mathbf{A}_{j-1} \frac{\partial \mathbf{T}_j}{\partial q_j} \dot{q}_j \tag{4.30}$$

$$\mathbf{C}_j = \dot{\mathbf{B}}_j = \ddot{\mathbf{A}}_j = \mathbf{C}_{j-1} \mathbf{T}_j + 2\mathbf{B}_{j-1} \frac{\partial \mathbf{T}_j}{\partial q_j} \dot{q}_j + \mathbf{A}_{j-1} \frac{\partial^2 \mathbf{T}_j}{\partial q_j{}^2} \dot{q}_j{}^2 + \mathbf{A}_{j-1} \frac{\partial \mathbf{T}_j}{\partial q_j} \ddot{q}_j \tag{4.31}$$

where $\mathbf{A}_0 = [\mathbf{I}]$ and $\mathbf{B}_0 = \mathbf{C}_0 = [\mathbf{0}]$.

After obtaining all the transformation matrices $\mathbf{A}_j$, $\mathbf{B}_j$, $\mathbf{C}_j$, the global position, velocity, and acceleration of a point in Cartesian coordinates can be calculated as

$$^0\mathbf{r}_j = \mathbf{A}_j \mathbf{r}_j; \, ^0\dot{\mathbf{r}}_j = \mathbf{B}_j \mathbf{r}_j; \, ^0\ddot{\mathbf{r}}_j = \mathbf{C}_j \mathbf{r}_j \tag{4.32}$$

where $\mathbf{r}_j$ contains the augmented local coordinates of the point in j^{th} coordinate system.

4.5.2 Backward recursive dynamics

Based on forward recursive kinematics, the backward recursion for dynamic analysis is accomplished by defining 4×4 transformation matrix $\mathbf{D}_i$ and 4×1 transformation matrices $\mathbf{E}_i$, $\mathbf{F}_i$, and $\mathbf{G}_i$ as follows.

Given the mass and inertia properties of each link, and the external force $\mathbf{f}_k^T = [^k f_x \quad ^k f_y \quad ^k f_z \quad 0]$ and the moment $\mathbf{h}_k^T = [^k h_x \quad ^k h_y \quad ^k h_z \quad 0]$ for the link k defined in the global coordinate system, then the joint actuation torques τ_i are computed for $i = n$ to 1 as (Hollerbach, 1980):

$$\tau_i = tr\left[\frac{\partial \mathbf{A}_i}{\partial q_i} \mathbf{D}_i\right] - \mathbf{g}^T \frac{\partial \mathbf{A}_i}{\partial q_i} \mathbf{E}_i - \mathbf{f}_k^T \frac{\partial \mathbf{A}_i}{\partial q_i} \mathbf{F}_i - \mathbf{G}_i^T \mathbf{A}_{i-1} \mathbf{z}_0 \tag{4.33}$$

where

$$\mathbf{D}_i = \mathbf{I}_i \mathbf{C}_i^T + \mathbf{T}_{i+1} \mathbf{D}_{i+1} \tag{4.34}$$

$$\mathbf{E}_i = m_i{}^i \mathbf{r}_i + \mathbf{T}_{i+1} \mathbf{E}_{i+1} \tag{4.35}$$

$$\mathbf{F}_i = {}^k \mathbf{r}_f \delta_{ik} + \mathbf{T}_{i+1} \mathbf{F}_{i+1} \tag{4.36}$$

$$\mathbf{G}_i = \mathbf{h}_k \delta_{ik} + \mathbf{G}_{i+1} \tag{4.37}$$

with $\mathbf{D}_{n+1} = \mathbf{E}_{n+1} = \mathbf{F}_{n+1} = \mathbf{G}_{n+1} = [\mathbf{0}]$; $\mathbf{I}_i$ is the inertia matrix for link i; m_i is the mass of link i; $\mathbf{g}$ is the gravity vector; $^i\mathbf{r}_i$ is the location of center of mass of link i in the local frame i; $^k\mathbf{r}_f$ is position of the external force in the local frame k; $\mathbf{z}_0 = [0 \quad 0 \quad 1 \quad 0]^T$ for a revolute joint and $\mathbf{z}_0 = [0 \quad 0 \quad 0 \quad 0]^T$ for a prismatic joint. δ_{ik} is Kronecker delta.

The first term in torque expression is the inertia and Coriolis torque, the second term denotes the torque of force due to gravity, the third term is the torque due to external force, and the fourth term represents the torque due to external moment.

4.5.3 Sensitivity analysis

The derivatives, $\dfrac{\partial \tau_i}{\partial q_k}$, $\dfrac{\partial \tau_i}{\partial \dot{q}_k}$, $\dfrac{\partial \tau_i}{\partial \ddot{q}_k}$ ($i = 1$ to n; $k = 1$ to n), can be evaluated for a mechanical system in a recursive way using the foregoing recursive Lagrangian dynamics formulation.

4.5.4 Kinematics sensitivity analysis

For a given point, sensitivity of position, velocity, and acceleration with respect to state variables relates to transformation matrices $\mathbf{A}$, $\mathbf{B}$, and $\mathbf{C}$:

$$\frac{\partial \mathbf{A}_i}{\partial q_k} = \begin{cases} \dfrac{\partial \mathbf{A}_{i-1}}{\partial q_k} \mathbf{T}_i & (k < i) \\[2ex] \mathbf{A}_{i-1} \dfrac{\partial \mathbf{T}_i}{\partial q_k} & (k = i) \\[2ex] \mathbf{0} & (k > i) \end{cases} \tag{4.38}$$

$$\frac{\partial \mathbf{B}_i}{\partial q_k} = \begin{cases} \dfrac{\partial \mathbf{B}_{i-1}}{\partial q_k} \mathbf{T}_i + \dfrac{\partial \mathbf{A}_{i-1}}{\partial q_k} \dfrac{\partial \mathbf{T}_i}{\partial q_i} \dot{q}_i & (k < i) \\[2ex] \mathbf{B}_{i-1} \dfrac{\partial \mathbf{T}_i}{\partial q_k} + \mathbf{A}_{n-1} \dfrac{\partial^2 \mathbf{T}_i}{\partial q_k^2} \dot{q}_i & (k = i) \\[2ex] \mathbf{0} & (k > i) \end{cases} \tag{4.39}$$

$$\frac{\partial \mathbf{B}_i}{\partial \dot{q}_k} = \begin{cases} \dfrac{\partial \mathbf{B}_{i-1}}{\partial \dot{q}_k} \mathbf{T}_i & (k < i) \\[2ex] \mathbf{A}_{i-1} \dfrac{\partial \mathbf{T}_i}{\partial q_k} & (k = i) \\[2ex] \mathbf{0} & (k > i) \end{cases} \tag{4.40}$$

$$\frac{\partial \mathbf{C}_i}{\partial q_k} = \begin{cases} \dfrac{\partial \mathbf{C}_{i-1}}{\partial q_k} \mathbf{T}_i + 2 \dfrac{\partial \mathbf{B}_{i-1}}{\partial q_k} \dfrac{\partial \mathbf{T}_i}{\partial q_i} \dot{q}_i + \dfrac{\partial \mathbf{A}_{i-1}}{\partial q_k} \dfrac{\partial^2 \mathbf{T}_i}{\partial q_i^2} \dot{q}_i^2 + \dfrac{\partial \mathbf{A}_{i-1}}{\partial q_k} \dfrac{\partial \mathbf{T}_i}{\partial q_i} \ddot{q}_i & (k < i) \\[2ex] \mathbf{C}_{i-1} \dfrac{\partial \mathbf{T}_i}{\partial q_k} + 2 \mathbf{B}_{i-1} \dfrac{\partial^2 \mathbf{T}_i}{\partial q_k^2} \dot{q}_i + \mathbf{A}_{i-1} \dfrac{\partial^3 \mathbf{T}_i}{\partial q_k^3} \dot{q}_i^2 + \mathbf{A}_{i-1} \dfrac{\partial^2 \mathbf{T}_i}{\partial q_k^2} \ddot{q}_i & (k = i) \\[2ex] \mathbf{0} & (k > i) \end{cases} \tag{4.41}$$

$$\frac{\partial \mathbf{C}_i}{\partial \dot{q}_k} = \begin{cases} \dfrac{\partial \mathbf{C}_{i-1}}{\partial \dot{q}_k}\mathbf{T}_i + 2\dfrac{\partial \mathbf{B}_{i-1}}{\partial \dot{q}_k}\dfrac{\partial \mathbf{T}_i}{\partial q_i}\dot{q}_i & (k < i) \\[2ex] 2\mathbf{B}_{i-1}\dfrac{\partial \mathbf{T}_i}{\partial q_k} + 2\mathbf{A}_{i-1}\dfrac{\partial^2 \mathbf{T}_i}{\partial q_k^2}\dot{q}_i & (k = i) \\[2ex] 0 & (k > i) \end{cases} \tag{4.42}$$

$$\frac{\partial \mathbf{C}_i}{\partial \ddot{q}_k} = \begin{cases} \dfrac{\partial \mathbf{C}_{i-1}}{\partial \ddot{q}_k}\mathbf{T}_i & (k < i) \\[2ex] \mathbf{A}_{i-1}\dfrac{\partial \mathbf{T}_i}{\partial q_k} & (k = i) \\[2ex] 0 & (k > i) \end{cases} \tag{4.43}$$

The forward recursive kinematics sensitivity equations are implemented as follows:

Input state variables $\mathbf{q}, \dot{\mathbf{q}}, \ddot{\mathbf{q}}$ and initial condition $\mathbf{A}[0] = \mathbf{I}$, $\mathbf{B}[0] = \mathbf{0}$, $\mathbf{C}[0] = \mathbf{0}$

Do $i = 1, n$
    (1) Obtain $q_i, \dot{q}_i, \ddot{q}_i$
    (2) Calculate and store $\mathbf{T}_i(q_i), \partial \mathbf{T}_i(q_i)/\partial q_i, \partial^2 \mathbf{T}_i(q_i)/\partial q_i^2, \partial^3 \mathbf{T}_i(q_i)/\partial q_i^3$
    (3) Calculate and store $\mathbf{A}_i$
    (4) Calculate and store $\mathbf{B}_i$
    (5) Calculate and store $\mathbf{C}_i$ (4.44)
        (a) Calculate and store $\partial \mathbf{A}_i/\partial q_j$
        (b) Calculate and store $\partial \mathbf{B}_i/\partial q_j, \partial \mathbf{B}_i/\partial \dot{q}_j$
        (c) Calculate and store $\partial \mathbf{C}_i/\partial q_j, \partial \mathbf{C}_i/\partial \dot{q}_j, \partial \mathbf{C}_i/\partial \ddot{q}_j$
EndDo

4.5.5 Dynamics sensitivity analysis

Sensitivity of the joint torque with respect to the state variables involves $\mathbf{D}$, $\mathbf{E}$, $\mathbf{F}$, and $\mathbf{G}$, which correspond to inertia and Coriolis, gravity, external force, and external moment, respectively.

$$\frac{\partial \mathbf{D}_i}{\partial q_k} = \begin{cases} \mathbf{I}_i \dfrac{\partial \mathbf{C}_i^T}{\partial q_k} + \mathbf{T}_{i+1}\dfrac{\partial \mathbf{D}_{i+1}}{\partial q_k} & (k \le i) \\[2ex] \dfrac{\partial \mathbf{T}_{i+1}}{\partial q_k}\mathbf{D}_{i+1} + \mathbf{T}_{i+1}\dfrac{\partial \mathbf{D}_{i+1}}{\partial q_k} & (k = i+1) \\[2ex] \mathbf{T}_{i+1}\dfrac{\partial \mathbf{D}_{i+1}}{\partial q_k} & (k > i+1) \end{cases} \tag{4.45}$$

$$
\frac{\partial \mathbf{D}_i}{\partial \dot{q}_k} =
\begin{cases}
\mathbf{I}_i \dfrac{\partial \mathbf{C}_i^T}{\partial \dot{q}_k} + \mathbf{T}_{i+1} \dfrac{\partial \mathbf{D}_{i+1}}{\partial \dot{q}_k} & (k \le i) \\[3mm]
\mathbf{T}_{i+1} \dfrac{\partial \mathbf{D}_{i+1}}{\partial \dot{q}_k} & (k > i)
\end{cases}
\tag{4.46}
$$

$$
\frac{\partial \mathbf{D}_i}{\partial \ddot{q}_k} =
\begin{cases}
\mathbf{I}_i \dfrac{\partial \mathbf{C}_i^T}{\partial \ddot{q}_k} + \mathbf{T}_{i+1} \dfrac{\partial \mathbf{D}_{i+1}}{\partial \ddot{q}_k} & (k \le i) \\[3mm]
\mathbf{T}_{i+1} \dfrac{\partial \mathbf{D}_{i+1}}{\partial \ddot{q}_k} & (k > i)
\end{cases}
\tag{4.47}
$$

$$
\frac{\partial \mathbf{E}_i}{\partial q_k} =
\begin{cases}
\mathbf{0} & (k \le i) \\[3mm]
\dfrac{\partial \mathbf{T}_{i+1}}{\partial q_k} \mathbf{E}_{i+1} + \mathbf{T}_{i+1} \dfrac{\partial \mathbf{E}_{i+1}}{\partial q_k} & (k = i+1) \\[3mm]
\mathbf{T}_{i+1} \dfrac{\partial \mathbf{E}_{i+1}}{\partial q_k} & (k > i+1)
\end{cases}
\tag{4.48}
$$

$$
\frac{\partial \mathbf{F}_i}{\partial q_k} =
\begin{cases}
\mathbf{0} & (k \le i) \\[3mm]
\dfrac{\partial \mathbf{T}_{i+1}}{\partial q_k} \mathbf{F}_{i+1} + \mathbf{T}_{i+1} \dfrac{\partial \mathbf{F}_{i+1}}{\partial q_k} & (k = i+1) \\[3mm]
\mathbf{T}_{i+1} \dfrac{\partial \mathbf{F}_{i+1}}{\partial q_k} & (k > i+1)
\end{cases}
\tag{4.49}
$$

$$
\frac{\partial \mathbf{G}_i}{\partial q_k} = \mathbf{0}
\tag{4.50}
$$

$$
\frac{\partial \tau_i}{\partial q_k} =
\begin{cases}
tr\left(\dfrac{\partial^2 \mathbf{A}_i}{\partial q_i \partial q_k} \mathbf{D}_i + \dfrac{\partial \mathbf{A}_i}{\partial q_i} \dfrac{\partial \mathbf{D}_i}{\partial q_k} \right) - \mathbf{g}^T \dfrac{\partial^2 \mathbf{A}_i}{\partial q_i \partial q_k} \mathbf{E}_i - \mathbf{f}^T \dfrac{\partial^2 \mathbf{A}_i}{\partial q_i \partial q_k} \mathbf{F}_i - \mathbf{G}_i^T \dfrac{\partial \mathbf{A}_{i-1}}{\partial q_k} \mathbf{z}_0 & (k \le i) \\[4mm]
tr\left(\dfrac{\partial \mathbf{A}_i}{\partial q_i} \dfrac{\partial \mathbf{D}_i}{\partial q_k} \right) - \mathbf{g}^T \dfrac{\partial \mathbf{A}_i}{\partial q_i} \dfrac{\partial \mathbf{E}_i}{\partial q_k} - \mathbf{f}^T \dfrac{\partial \mathbf{A}_i}{\partial q_i} \dfrac{\partial \mathbf{F}_i}{\partial q_k} & (k > i)
\end{cases}
\tag{4.51}
$$

$$
\frac{\partial \tau_i}{\partial \dot{q}_k} = tr\left(\frac{\partial \mathbf{A}_i}{\partial q_i} \frac{\partial \mathbf{D}_i}{\partial \dot{q}_k} \right)
\tag{4.52}
$$

$$
\frac{\partial \tau_i}{\partial \ddot{q}_k} = tr\left(\frac{\partial \mathbf{A}_i}{\partial q_i} \frac{\partial \mathbf{D}_i}{\partial \ddot{q}_k} \right)
\tag{4.53}
$$

Thus, the gradients of torque with respect to state variables are obtained through Equations (4.51−4.53). It is essential to have closed-form expressions for the gradients as this facilitates fast computation of the optimization algorithm.

The backward recursive dynamics sensitivity equations are implemented as follows:

Input state variables $\mathbf{q}, \dot{\mathbf{q}}, \ddot{\mathbf{q}}$ and final condition $\mathbf{D}[n+1] = \mathbf{0}$, $\mathbf{E}[n+1] = \mathbf{0}$, $\mathbf{F}[n+1] = \mathbf{0}$, $\mathbf{G}[n+1] = \mathbf{0}$ and use the stored information calculated from forward recursive sensitivity algorithm.

$$
\begin{aligned}
&\textbf{Do } i = n, 1 \\
&\quad \text{(1) Calculate and store } \mathbf{D}_i, \mathbf{E}_i, \mathbf{F}_i, \mathbf{G}_i \\
&\quad \text{(2) Calculate and store } \tau_i \\
&\quad \textbf{Do } j = 1, i \\
&\quad\quad \text{Calculate and store } \partial^2 A_i / \partial q_i \partial q_j \\
&\quad \textbf{End Do}
\end{aligned}
\tag{4.54}
$$

$$
\begin{aligned}
&\textbf{Do } k = 1, n \\
&\quad \text{(a) Calculate and store } \partial \mathbf{D}_i / \partial q_k, \partial \mathbf{D}_i / \partial \dot{q}_k, \partial \mathbf{D}_i / \partial \ddot{q}_k \\
&\quad \text{(b) Calculate and store } \partial \mathbf{E}_i / \partial q_k \\
&\quad \text{(c) Calculate and store } \partial \mathbf{F}_i / \partial q_k \\
&\quad \text{(d) Calculate and store } \partial \tau_i / \partial q_k, \partial \tau_i / \partial \dot{q}_k, \partial \tau_i / \partial \ddot{q}_k \\
&\quad \textbf{EndDo} \\
&\textbf{EndDo}
\end{aligned}
\tag{4.55}
$$

4.5.6 Joint profile discretization

A joint profile $q(t)$ is parameterized by using uniform B-splines as follows:

$$
q(\mathbf{t}, \mathbf{P}) = \sum_{i=1}^{m} B_i(\mathbf{t}) p_i \quad 0 \leq t \leq T
\tag{4.56}
$$

where $B_i(\mathbf{t})$ are the basis functions, $\mathbf{t} = \{t_0, \ldots, t_s\}$ is the knot vector, and $\mathbf{P} = \{p_1, \ldots, p_m\}$ is the control point vector. With this representation, the control points become the optimization variables (also called the design variables). B-spline interpolation has many important properties, such as continuity, differentiability, and local control. These properties, especially differentiability and local control, make B-splines competent to represent joint angle trajectories, which require smoothness and flexibility (Wang et al., 2007).

The B-spline basis functions are uniquely determined by knot vector $\mathbf{t}$, which is evenly spaced on the time interval $[0 \quad T]$ with time step Δt, as follows:

$$
t_{i+1} = t_i + \Delta t, \quad \Delta t = \frac{T}{s}, \quad i = 0, \ldots, s-1
\tag{4.57}
$$

where s is the number of discretized segments.

Note that q, $\dot{q}$, and $\ddot{q}$ are calculated as functions of $\mathbf{t}$ and $\mathbf{P}$; therefore torque $\tau = \tau(\mathbf{t}, \mathbf{P})$ is an explicit function of the knot vector and control points from the equation of motion. Thus, the derivatives of a torque τ with respect to the control points and knot points can be computed using the chain rule as:

$$\frac{\partial \tau}{\partial p_i} = \frac{\partial \tau}{\partial q}\frac{\partial q}{\partial p_i} + \frac{\partial \tau}{\partial \dot{q}}\frac{\partial \dot{q}}{\partial p_i} + \frac{\partial \tau}{\partial \ddot{q}}\frac{\partial \ddot{q}}{\partial p_i} \tag{4.58}$$

$$\frac{\partial \tau}{\partial t_i} = \frac{\partial \tau}{\partial q}\frac{\partial q}{\partial t_i} + \frac{\partial \tau}{\partial \dot{q}}\frac{\partial \dot{q}}{\partial t_i} + \frac{\partial \tau}{\partial \ddot{q}}\frac{\partial \ddot{q}}{\partial t_i} \tag{4.59}$$

4.6 Examples using a 2-DOF arm

We first derive the recursive Lagrangian sensitivity equations for the 2-DOF system. We then use these derivations in numerical examples to illustrate the use of the Lagrangian recursive formulations in solving general-purpose problems.

The first problem is the time-optimal trajectory-planning design without gravity and external forces. The reason for using this example is to validate the numerical results, as these solutions have been well studied and are readily available in the literature (Dissanayake et al., 1991; Wang et al., 2005).

The second problem is to optimize a lifting motion of the arm with a mixed performance criterion. Both gravity and external force are considered. Sensitivity results with the recursive algorithm and the closed-form formulation are numerically compared.

For the purpose of demonstrating the formulation for dynamics, consider a 2-DOF arm constrained to move in the vertical plane—shown in Figure 4.2 (vertical motion only to reduce the complexity for this example while taking gravity into consideration). We shall first derive the recursive Lagrangian sensitivity equations for the arm. This system consists of two links whose lengths are L_1 and L_2, and moments of inertia are I_1 and I_2 as shown in Figure 4.2. The relative joint angles are denoted by q_1 and q_2, respectively, and are controlled by the joint actuating torques τ_1 and τ_2. The two segmental links of the arm are considered rigid, and the relative joint angles are selected as independent generalized coordinates. The system is assumed to lie in the vertical plane restricting the motion to 2 DOFs. Actuator torques (muscle actions) drive the arm from the initial position $(q_1(0), q_2(0))$ to the final position $(q_1(T), q_2(T))$ in the time interval T. In addition, the arm is at rest at the initial and final points. The data for the arm are given in Table 4.1.

This arm can be modeled as a planar kinematic chain as shown in Figure 4.3, where l_1 and l_2 are the distances from the local coordinate system to the center of mass location.

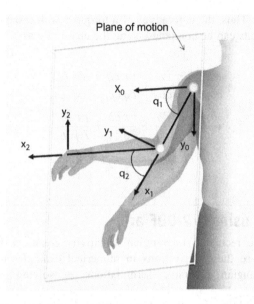

Plane of motion

FIGURE 4.2

A 2-DOF planar arm model restricted to move in a planar motion and showing the DH coordinate systems.

Table 4.1 Parameters for the 2-DOF Arm

Length:	$L_1 = L_2 = 0.4\,m$
Mass:	$m_1 = m_2 = 0.5\,kg$
Inertia:	$I_1 = I_2 = 0.1\,kgm^2$
Center of Mass of Link 1:	$^1r_1 = (-L_1/2, 0)$
Center of Mass of Link 2:	$^2r_2 = (-L_2/2, 0)$
Torque upper bound:	$\tau^U = 10\,Nm$
Torque lower bound:	$\tau^L = -10\,Nm$

4.6.1 The DH parameters

The DH parameters of the planar two-link arm are shown in Figure 4.3 and presented in Table 4.2.

The transformation matrices are:

$$T_1 = \begin{pmatrix} cos\theta_1 & -sin\theta_1 & 0 & L_1cos\theta_1 \\ sin\theta_1 & cos\theta_1 & 0 & L_1sin\theta_1 \\ 0 & 0 & 1 & 0 \\ 0 & 0 & 0 & 1 \end{pmatrix} \quad T_2 = \begin{pmatrix} cos\theta_2 & -sin\theta_2 & 0 & L_2cos\theta_2 \\ sin\theta_2 & cos\theta_2 & 0 & L_2sin\theta_2 \\ 0 & 0 & 1 & 0 \\ 0 & 0 & 0 & 1 \end{pmatrix} \quad (4.60)$$

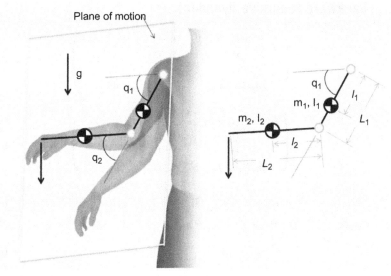

FIGURE 4.3

A 2-DOF kinematic skeleton of an arm restricted to move on a vertical plane and showing the center of mass locations and dimensions.

Table 4.2 DH parameters for the Two-link Arm

Link	θ_i	d_i	a_i	α_i
1	θ_1	0	L_1	0
2	θ_2	0	L_2	0

4.6.2 Forward recursive kinematics

In order to obtain expressions for the forward recursive kinematics, we first find expressions for the A, B, and C terms.

$$A_1 = T_1; \quad A_2 = A_1 T_2 \tag{4.61}$$

$$B_1 = \frac{\partial T_1}{\partial \theta_1} \dot{\theta}_1; \quad B_2 = B_1 T_2 + A_1 \frac{\partial T_2}{\partial \theta_2} \dot{\theta}_2 \tag{4.62}$$

$$C_1 = \frac{\partial^2 T_1}{\partial \theta_1^2} \dot{\theta}_1^2 + \frac{\partial T_1}{\partial \theta_1} \ddot{\theta}_1; \quad C_2 = C_1 T_2 + 2B_1 \frac{\partial T_2}{\partial \theta_2} \dot{\theta}_2 + A_1 \frac{\partial^2 T_2}{\partial \theta_2^2} \dot{\theta}_2^2 + A_1 \frac{\partial T_2}{\partial \theta_2} \ddot{\theta}_2 \tag{4.63}$$

4.6.3 Backward recursive dynamics

$$J_2 = \begin{pmatrix} I_2 + m_2(l_2 - L_2)^2 & 0 & 0 & m_2(l_2 - L_2) \\ 0 & 0 & 0 & 0 \\ 0 & 0 & 0 & 0 \\ m_2(l_2 - L_2) & 0 & 0 & m_2 \end{pmatrix} \qquad (4.64)$$

$$J_1 = \begin{pmatrix} I_1 + m_1(l_1 - L_1)^2 & 0 & 0 & m_1(l_1 - L_1) \\ 0 & 0 & 0 & 0 \\ 0 & 0 & 0 & 0 \\ m_1(l_1 - L_1) & 0 & 0 & m_1 \end{pmatrix} \qquad (4.65)$$

$$D_2 = J_2 C_2^T; \quad D_1 = J_1 C_1^T + T_2 D_2 \qquad (4.66)$$

$$E_2 = m_2 (l_2 - L_2 \ \ 0 \ \ 0 \ \ 1)^T; \quad E_1 = m_1(l_1 - L_1 \ \ 0 \ \ 0 \ \ 1)^T + T_2 E_2 \qquad (4.67)$$

$$F_2 = (0 \ \ 0 \ \ 0 \ \ 1)^T \quad F_1 = T_2 F_2 \qquad (4.68)$$

$$\mathbf{g} = (0 \ \ -g \ \ 0 \ \ 0)^T \quad \mathbf{f}_2 = (0 \ \ -f \ \ 0 \ \ 0)^T \qquad (4.69)$$

And the torques are expressed as:

$$\tau_1 = tr \left[\frac{\partial \mathbf{A}_1}{\partial q_1} \mathbf{D}_1 \right] - \mathbf{g}^T \frac{\partial \mathbf{A}_1}{\partial q_1} \mathbf{E}_1 - \mathbf{f}_2^T \frac{\partial \mathbf{A}_1}{\partial q_1} \mathbf{F}_1$$

$$= (I_1 + I_2 + m_1 l_1^2 + m_2(L_1^2 + l_2^2 + 2L_1 l_2 \cos \theta_2)) \ddot{\theta}_1 + (I_2 + m_2 l_2^2 + m_2 L_1 l_2 \cos \theta_2) \ddot{\theta}_2$$
$$- 2m_2 L_1 l_2 \dot{\theta}_1 \dot{\theta}_2 \sin \theta_2 - m_2 L_1 l_2 \dot{\theta}_2^2 \sin \theta_2 + m_2 g l_2 \cos(\theta_1 + \theta_2) + m_1 g l_1 \cos \theta_1$$
$$+ m_2 g L_1 \cos \theta_1 + f L_2 \cos(\theta_1 + \theta_2) + f L_1 \cos \theta_1$$

$$(4.70)$$

$$\tau_2 = tr \left[\frac{\partial \mathbf{A}_2}{\partial q_2} \mathbf{D}_2 \right] - \mathbf{g}^T \frac{\partial \mathbf{A}_2}{\partial q_2} \mathbf{E}_2 - \mathbf{f}_2^T \frac{\partial \mathbf{A}_2}{\partial q_2} \mathbf{F}_2$$

$$(4.71)$$

$$= (I_2 + m_2 l_2^2) \ddot{\theta}_2 + (I_2 + m_2 l_2^2 + m_2 L_1 l_2 \cos \theta_2) \ddot{\theta}_1 + m_2 L_1 l_2 \dot{\theta}_1^2 \sin \theta_2$$
$$+ m_2 g l_2 \cos(\theta_1 + \theta_2) + f L_2 \cos(\theta_1 + \theta_2)$$

4.6.4 Gradients

Explicit gradients of torque with respect to state variables are derived as follows:

$$\frac{\partial \tau_1}{\partial \theta_1} = tr \left(\frac{\partial^2 \mathbf{A}_1}{\partial q_1 \partial q_1} \mathbf{D}_1 + \frac{\partial \mathbf{A}_1}{\partial q_1} \frac{\partial \mathbf{D}_1}{\partial q_1} \right) - \mathbf{g}^T \frac{\partial^2 \mathbf{A}_1}{\partial q_1 \partial q_1} \mathbf{E}_1 - \mathbf{f}_2^T \frac{\partial^2 \mathbf{A}_1}{\partial q_1 \partial q_1} \mathbf{F}_1$$

$$= - m_2 g l_2 \sin(\theta_1 + \theta_2) - m_1 g l_1 \sin \theta_1 - m_2 g L_1 \sin \theta_1 - f L_2 \sin(\theta_1 + \theta_2) - f L_1 \sin \theta_1$$

$$(4.72)$$

$$\frac{\partial \tau_1}{\partial \theta_2} = tr\left(\frac{\partial \mathbf{A}_1}{\partial q_1}\frac{\partial \mathbf{D}_1}{\partial q_2}\right) - \mathbf{g}^T\frac{\partial \mathbf{A}_1}{\partial q_1}\frac{\partial \mathbf{E}_1}{\partial q_2} - \mathbf{f}_2^T\frac{\partial \mathbf{A}_1}{\partial q_1}\frac{\partial \mathbf{F}_1}{\partial q_2}$$

$$= (-2m_2L_1l_2\sin\theta_2)\ddot{\theta}_1 + (-m_2L_1l_2\sin\theta_2)\ddot{\theta}_2 - 2m_2L_1l_2\dot{\theta}_1\dot{\theta}_2\cos\theta_2$$

$$- m_2L_1l_2\dot{\theta}_2^2\cos\theta_2 - m_2gl_2\sin(\theta_1+\theta_2) - fL_2\sin(\theta_1+\theta_2) \tag{4.73}$$

$$\frac{\partial \tau_1}{\partial \dot{\theta}_1} = tr\left(\frac{\partial \mathbf{A}_1}{\partial q_1}\frac{\partial \mathbf{D}_1}{\partial \dot{\theta}_1}\right) = -2m_2L_1l_2\dot{\theta}_2\sin\theta_2 \tag{4.74}$$

$$\frac{\partial \tau_1}{\partial \dot{\theta}_2} = tr\left(\frac{\partial \mathbf{A}_1}{\partial q_1}\frac{\partial \mathbf{D}_1}{\partial \dot{\theta}_2}\right) = -2m_2L_1l_2\dot{\theta}_1\sin\theta_2 - 2m_2L_1l_2\dot{\theta}_2\sin\theta_2 \tag{4.75}$$

$$\frac{\partial \tau_1}{\partial \ddot{\theta}_1} = tr\left(\frac{\partial \mathbf{A}_1}{\partial q_1}\frac{\partial \mathbf{D}_1}{\partial \ddot{\theta}_1}\right) = I_1 + I_2 + m_1l_1^2 + m_2(L_1^2 + l_2^2 + 2L_1l_2\cos\theta_2) \tag{4.76}$$

$$\frac{\partial \tau_1}{\partial \ddot{\theta}_2} = tr\left(\frac{\partial \mathbf{A}_1}{\partial q_1}\frac{\partial \mathbf{D}_1}{\partial \ddot{\theta}_2}\right) = I_2 + m_2l_2^2 + m_2L_1l_2\cos\theta_2 \tag{4.77}$$

$$\frac{\partial \tau_2}{\partial \theta_1} = tr\left(\frac{\partial^2 \mathbf{A}_2}{\partial q_2\partial q_1}\mathbf{D}_2 + \frac{\partial \mathbf{A}_2}{\partial q_2}\frac{\partial \mathbf{D}_2}{\partial q_1}\right) - \mathbf{g}^T\frac{\partial^2 \mathbf{A}_2}{\partial q_2\partial q_1}\mathbf{E}_2 - \mathbf{f}_2^T\frac{\partial^2 \mathbf{A}_2}{\partial q_2\partial q_1}\mathbf{F}_2$$

$$= -m_2gl_2\sin(\theta_1+\theta_2) - fL_2\sin(\theta_1+\theta_2) \tag{4.78}$$

$$\frac{\partial \tau_2}{\partial \theta_2} = tr\left(\frac{\partial^2 \mathbf{A}_2}{\partial q_2\partial q_2}\mathbf{D}_2 + \frac{\partial \mathbf{A}_2}{\partial q_2}\frac{\partial \mathbf{D}_2}{\partial q_2}\right) - \mathbf{g}^T\frac{\partial^2 \mathbf{A}_2}{\partial q_2\partial q_2}\mathbf{E}_2 - \mathbf{f}_2^T\frac{\partial^2 \mathbf{A}_2}{\partial q_2\partial q_2}\mathbf{F}_2$$

$$= (-m_2L_1l_2\sin\theta_2)\ddot{\theta}_1 + m_2L_1l_2\dot{\theta}_1^2\cos\theta_2 - m_2gl_2\sin(\theta_1+\theta_2) - fL_2\sin(\theta_1+\theta_2)$$

$$\tag{4.79}$$

$$\frac{\partial \tau_2}{\partial \dot{\theta}_1} = tr\left(\frac{\partial \mathbf{A}_2}{\partial q_2}\frac{\partial \mathbf{D}_2}{\partial \dot{\theta}_1}\right) = 2m_2L_1l_2\dot{\theta}_1\sin\theta_2 \tag{4.80}$$

$$\frac{\partial \tau_2}{\partial \dot{\theta}_2} = tr\left(\frac{\partial \mathbf{A}_2}{\partial q_2}\frac{\partial \mathbf{D}_2}{\partial \dot{\theta}_2}\right) = 0 \tag{4.81}$$

$$\frac{\partial \tau_2}{\partial \ddot{\theta}_1} = tr\left(\frac{\partial \mathbf{A}_2}{\partial q_2}\frac{\partial \mathbf{D}_2}{\partial \ddot{\theta}_1}\right) = I_2 + m_2l_2^2 + m_2L_1l_2\cos\theta_2 \tag{4.82}$$

$$\frac{\partial \tau_2}{\partial \ddot{\theta}_2} = tr\left(\frac{\partial \mathbf{A}_2}{\partial q_2}\frac{\partial \mathbf{D}_2}{\partial \ddot{\theta}_2}\right) = I_2 + m_2l_2^2 \tag{4.83}$$

The foregoing recursive sensitivity equations can be readily verified with the closed-form solution for this problem as detailed below.

4.6.5 Closed-form equations of motion

In order to compare the Lagrangian formulation with the closed form, we will derive below the closed-form equations. We shall impose gravity g and a force f at the end-effector.

The closed-form Lagrangian equation of a two-link rigid arm is well studied and can be written as follows:

$$\tau_1 = (I_1 + I_2 + m_1 l_1^2 + m_2(L_1^2 + l_2^2 + 2L_1 l_2 \cos\theta_2))\ddot{\theta}_1 + (I_2 + m_2 l_2^2 + m_2 L_1 l_2 \cos\theta_2)\ddot{\theta}_2$$
$$- 2m_2 L_1 l_2 \dot{\theta}_1 \dot{\theta}_2 \sin\theta_2 - m_2 L_1 l_2 \dot{\theta}_2^2 \sin\theta_2 + m_2 g l_2 \cos(\theta_1 + \theta_2) + m_1 g l_1 \cos\theta_1$$
$$+ m_2 g L_1 \cos\theta_1 + fL_2 \cos(\theta_1 + \theta_2) + fL_1 \cos\theta_1$$

$$(4.84)$$

$$\tau_2 = (I_2 + m_2 l_2^2)\ddot{\theta}_2 + (I_2 + m_2 l_2^2 + m_2 L_1 l_2 \cos\theta_2)\ddot{\theta}_1 + m_2 L_1 l_2 \dot{\theta}_1^2 \sin\theta_2$$
$$+ m_2 g l_2 \cos(\theta_1 + \theta_2) + fL_2 \cos(\theta_1 + \theta_2)$$

$$(4.85)$$

Explicit gradients of torque with respect to state variables are derived as follows:

$$\frac{\partial \tau_1}{\partial \theta_1} = - m_2 g l_2 \sin(\theta_1 + \theta_2) - m_1 g l_1 \sin\theta_1 - m_2 g L_1 \sin\theta_1 - fL_2 \sin(\theta_1 + \theta_2) - fL_1 \sin\theta_1$$

$$(4.86)$$

$$\frac{\partial \tau_1}{\partial \theta_2} = (- 2m_2 L_1 l_2 \sin\theta_2)\ddot{\theta}_1 + (- m_2 L_1 l_2 \sin\theta_2)\ddot{\theta}_2 - 2m_2 L_1 l_2 \dot{\theta}_1 \dot{\theta}_2 \cos\theta_2$$

$$(4.87)$$

$$- m_2 L_1 l_2 \dot{\theta}_2^2 \cos\theta_2 - m_2 g l_2 \sin(\theta_1 + \theta_2) - fL_2 \sin(\theta_1 + \theta_2)$$

$$\frac{\partial \tau_1}{\partial \dot{\theta}_1} = - 2m_2 L_1 l_2 \dot{\theta}_2 \sin\theta_2 \qquad (4.88)$$

$$\frac{\partial \tau_1}{\partial \dot{\theta}_2} = - 2m_2 L_1 l_2 \dot{\theta}_1 \sin\theta_2 - 2m_2 L_1 l_2 \dot{\theta}_2 \sin\theta_2 \qquad (4.89)$$

$$\frac{\partial \tau_1}{\partial \ddot{\theta}_1} = I_1 + I_2 + m_1 l_1^2 + m_2(L_1^2 + l_2^2 + 2L_1 l_2 \cos\theta_2) \qquad (4.90)$$

$$\frac{\partial \tau_1}{\partial \ddot{\theta}_2} = I_2 + m_2 l_2^2 + m_2 L_1 l_2 \cos\theta_2 \qquad (4.91)$$

$$\frac{\partial \tau_2}{\partial \theta_1} = - m_2 g l_2 \sin(\theta_1 + \theta_2) - fL_2 \sin(\theta_1 + \theta_2) \qquad (4.92)$$

$$\frac{\partial \tau_2}{\partial \theta_2} = (- m_2 L_1 l_2 \sin\theta_2)\ddot{\theta}_1 + m_2 L_1 l_2 \dot{\theta}_1^2 \cos\theta_2 - m_2 g l_2 \sin(\theta_1 + \theta_2) - fL_2 \sin(\theta_1 + \theta_2)$$

$$(4.93)$$

$$\frac{\partial \tau_2}{\partial \dot{\theta}_1} = 2m_2 L_1 l_2 \dot{\theta}_1 \sin \theta_2 \qquad (4.94)$$

$$\frac{\partial \tau_2}{\partial \dot{\theta}_2} = 0 \qquad (4.95)$$

$$\frac{\partial \tau_2}{\partial \ddot{\theta}_1} = I_2 + m_2 l_2^2 + m_2 L_1 l_2 \cos \theta_2 \qquad (4.96)$$

$$\frac{\partial \tau_2}{\partial \ddot{\theta}_2} = I_2 + m_2 l_2^2 \qquad (4.97)$$

The gradients will become import in later chapters when the optimization formulation is implemented.

4.7 **Trajectory planning example**

As an illustration of the use of recursive dynamics, we shall present an example of trajectory planning. Trajectory planning is defined as creating motion for an end-effector (e.g., a hand) from one point to another while avoiding collisions. We shall use concepts from Chapters 3 on optimization and Chapter 4 on dynamics to present a time-optimal trajectory planning for the two-link arm treated earlier. We shall solve the optimization problem using the recursive Lagrangian formulation.

The objective is to minimize total travel time T subjected to boundary conditions and torque limits. The same problem has been examined by Dissanayake et al. (1991) and Wang et al. (2005) using the closed-form equation of motion without the gravity effects. The parameter optimization problem can then be stated mathematically as follows: to compute design variables x, which are control points $\mathbf{P}$ and total travel time T, and to minimize T subject to the constraints on boundary conditions and torque limits

$$
\begin{aligned}
&Minimize. \quad T(\mathbf{x}) \\
&Such \ that \quad q_1(0) = 0.0, q_2(0) = -2.0 \\
&\qquad\qquad\ q_1(T) = 1.0, q_2(T) = -1.0 \\
&\qquad\qquad\ \dot{q}_1(0) = \dot{q}_2(0) = \dot{q}_1(T) = \dot{q}_2(T) = 0.0 \\
&\qquad\qquad\ -10 \le \tau(\mathbf{x}) \le 10
\end{aligned} \qquad (4.98)
$$

where T is the final time that needs to be minimized. Gravity effects are neglected in this time-optimal design problem, i.e., $g = 0$.

We use the word profile to denote a quantity changing over time. The optimal joint profiles are shown in Figure 4.4 and the joint torque profiles are depicted in Figure 4.5.

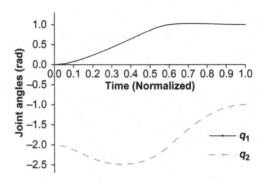

FIGURE 4.4

Joint angle profiles.

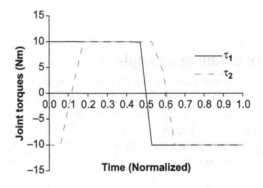

FIGURE 4.5

Joint torque profiles.

The minimum travel time is $T = 0.3934$ s, which is similar to that reported in the literature (0.3945 s by Furukawa [2002]; 0.394 s by Wang et al. [2005]). The optimality and feasibility tolerances are both set to the default value $\varepsilon = 10^{-6}$ in SNOPT. Different starting points were tried for the optimization and they all converge to the same optimal solution, implying a global solution for the problem. The convergence history of the cost function (the total travel time T) is plotted in Figure 4.6. It is noted that the SQP algorithm in SNOPT worked quite well and optimal solutions were obtained in 1.07 CPU seconds on a 1.40-GHz PC.

4.8 Arm lifting motion with load example

In this section, we shall use a simple introductory example to illustrate the capability of this method to address the lifting motion of a load with the arm.

The lifting motion of the two-link arm is studied by using a simple optimization problem where we shall use an energy function as a human performance

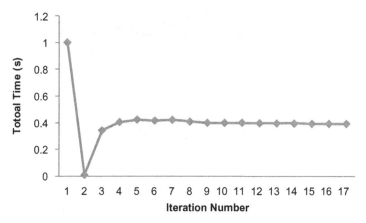

FIGURE 4.6

Iteration history for the cost function.

measure. This cost function will be optimized to predict the motion. Note that this example serves as an introduction to the following chapter, "Predictive dynamics," which is the essence of this book.

Consider an energy consumption cost function defined as

$$\int_{t=0}^{T} \left((1-u) + u \sum_{i=1}^{n} \tau_i^2 \right) dt \quad u \in [0,1] \tag{4.99}$$

where u is a specified constant and T is the total travel time of the arm. The second term, torque square, is related to energy consumption. When u goes to zero, this cost function becomes the time-optimal criterion. An appropriate u avoids harsh functioning of the actuator torques (Bessonnet and Lallemand, 1990). We shall also add an external load to the arm as if it is carrying a load from one point to another.

We shall use the same example treated earlier of the two-link arm with physical parameters as listed in Table 4.1. The constant vertical external load $f = 10$ N acting at the hand and the gravity effect $g = 9.8062\ m/s^2$ are considered in this case. The optimization formulation is given as follows:

$$
\begin{aligned}
Min. \quad & \int_{t=0}^{T} \left((1-u) + u \sum_{i=1}^{n} \tau_i^2(x) \right) dt \quad u \in [0,1] \\
St. \quad & q_1(0) = 1.32, q_2(0) = -2.37 \\
& q_1(T) = 2.80, q_2(T) = -2.37 \\
& \dot{q}_1(0) = \dot{q}_2(0) = \dot{q}_1(T) = \dot{q}_2(T) = 0.0 \\
& -10 \le \tau(x) \le 10
\end{aligned}
\tag{4.100}
$$

where x is a vector of control points **P** (joint profile) and total time T: $x = [P, T]$. $u = 0.01$. Starting points are obtained by linear interpolation between the initial and final joint angles.

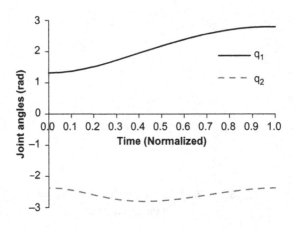

FIGURE 4.7

Joint angle profile with external load.

Note that the objective here is to predict the motion, which means predicting values for the joints as time progresses, also called joint profiles. The joint profiles are shown as the continuous and the dotted lines in the graph in Figure 4.7.

Because of the recursive dynamics approach with optimization, it is also possible to predict the torques as a function of time (also called torque profiles as shown in Figure 4.8), for the entire lifting motion.

In this example, we have used a numerical optimization solver called SNOPT. The optimality and feasibility tolerances are both set to the default value $\varepsilon = 10^{-6}$ in SNOPT and the optimal solutions were obtained in 2.44 CPU seconds on a 1.40-GHz PC. Optimal travel time for the mixed optimization problem is $T = 0.520$ s. To verify the optimal solution, the problem was also solved using commercial multi-body dynamics software called ADAMS™. The optimized joint torques as shown in Figure 4.9 were treated as inputs and the equations of motion were automatically generated and integrated to obtain the response. The two motion trajectories matched quite well. Sensitivity results with the recursive algorithm and the closed-form formulation at the optimal design are compared in Figure 4.9 and Figure 4.10. The sensitivity obtained from the two algorithms match quite closely.

4.9 Concluding remarks

Dynamics of human limbs are best modeled using the DH parameters and the Lagrangian formulation. This methodology is often chosen because it presents an elegant and systematic method for representing the motion including all aspects of dynamics, particularly for a multi-body serial system such as the human body. It is also readily suitable for computer implementation.

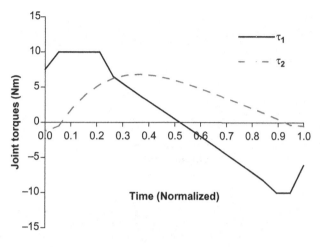

FIGURE 4.8

Joint torque profile with external load.

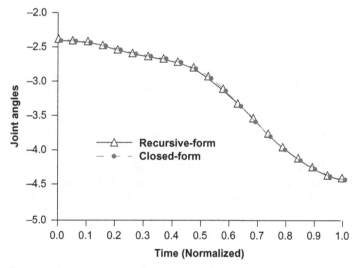

FIGURE 4.9

Sensitivity results compared for both the recursive and closed form.

It is essential to understand that the results of the recursive Lagrangian equations are indeed identical to those obtained from the closed-form equations of motion. In human modeling and simulation, particularly when a high degree of fidelity model is sought with a large number of DOF, the recursive Lagrangian formulation lends itself to numerical solutions.

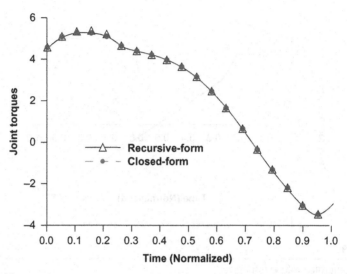

FIGURE 4.10

Sensitivity results of the joint torques comparison at the optimal design.

External forces and moments are systematically included in the recursive formulation. Although recursive, Newton−Euler also gives linear computational complexity with respect to the number of DOFs; recursive Lagrangian formulation is of higher efficiency and simplicity in calculating joint torques since the internal force calculation is avoided.

The examples of trajectory planning and the arm lifting motion of a model of a 2-DOF arm were solved for motion prediction. These simple examples illustrate the power and potential of using optimization with dynamics to "predict" human motion, as will be explored more in the next chapter.

In later chapters we will use the same idea with more complex motions, such as walking and lifting, for a full model of a human with substantially increased DOFs.

References

Bessonnet, G., Lallemand, J. P., 1990. Optimal trajectories of robot arms minimizing constrained actuators and travelling time. In: Robotics and Automation, 1990. Proceedings, 1990 IEEE International Conference. pp. 112−117.

Dissanayake, M.W.M.G., Goh, C.J., Phan-Thien, N., 1991. Time-optimal trajectories for robot manipulators. Robotica 9 (02), 131−138.

Fu, K.S., Gonzalez, R.C., Lee, C.S., 1987. Robotics. McGraw-Hill Book.

Furukawa, T., 2002. Time-subminimal trajectory planning for discrete non-linear systems. Eng. Optimiz. 34 (3), 219−243.

Hollerbach, J.M., 1980. A recursive Lagrangian formulation of manipulator dynamics and a comparative study of dynamics formulation complexity. IEEE T. Syst. Man Cyb. 10 (11), 730–736.

Sciavicco, L., Siciliano, B., 2000. Modelling and Control of Robot Manipulators. Springer Verlag.

Wang, Q., Xiang, Y., Kim, H., Arora, J. and Abdel-Malek, K. (2005). Alternative formulations for optimization-based digital human motion prediction (SAE Paper No. 05DHM-61). Warrendale, PA: SAE International.

Wang, Q., Xiang, Y., Arora, J. S. and Abdel-Malek, K. (2007). Alternative formulations for optimization-based human gait planning. 48th AIAA/ASME/ASCE/AHS/ASC Structures, Structural Dynamics and Materials Conference, Honolulu, Hawaii, Apr. 23–26.

Xiang, Y., Arora, J.S, Rahmatalla, S., Abdel-Malek, K., 2009a. Optimization-based dynamic human walking prediction: one step formulation. Int. J. Numer. Methods Eng. 79 (6), 667–695.

Xiang, Y., Arora, J.S., Rahmatalla, S., Bhatt, R., Marler, T., Abdel-Malek, K., 2009b. Human lifting simulation using multi-objective optimization approach. Multibody Syst. Dyn. 23 (4), 431–451.

Xiang, Y., Arora, J.S., Abdel-Malek, K., 2010a. Optimization-based prediction of asymmetric human gait. J. Biomech. 44 (4), 683–693.

Xiang, Y., Arora, J.S., Abdel-Malek, K., 2010b. Physics-based modeling and simulation of human walking: a review of optimization-based and other approaches. Struct. Multidiscip. Optim. 42 (1), 1–23.

Xiang, Y., Chung, H.J., Kim, J.H., Bhatt, R., Rahmatalla, S., Yang, J., et al., 2010c. Predictive dynamics: an optimization-based novel approach for human motion simulation. Struct. Multidiscip. Optim. 41 (3), 465–479.

Predictive Dynamics

Only two things are infinite, the universe and human stupidity, and I'm not sure about the former.
Albert Einstein (1879–1955)

5.1 Introduction

Predictive dynamics is a term coined to represent the methodology for predicting human motion while taking into consideration the biomechanics of the human and the physics of the task. It is *dynamics* in the sense that it deals with the equations of motion. It is *predictive* in that it is concerned with simulating or calculating what a human would do under the same conditions. At the heart of predictive dynamics is the cause and effect.

Predictive dynamics (PD) is an optimization-based approach for human motion prediction with unknown generalized forces and joint angle profiles. The basic idea is to model a redundant dynamical system as an optimization problem and to solve for its motion while optimizing a performance measure and satisfying the physical and kinematical constraints. Both the motion and the forces that cause the motion are unknown in the equations of motion that are treated as equality constraints and evaluated using inverse dynamics instead of their numerical integration. Available experimental data, such as response at discrete time points, can be used as constraints in the optimization formulation (Arora and Wang, 2005; Kim et al., 2005; Wang et al., 2005; Xiang et al., 2007; Xiang, 2008; Xiang et al., 2010a–c). This chapter is an adaptation of the work by Xiang et al. (2009c, 2010c).

5.2 Problem formulation

The basic idea of predictive dynamics is to model a redundant dynamical system as an optimization problem and to solve for its motion where only limited information about the system is available. It uses cost functions, which represent human performance measures, to drive the motion. The predictive dynamics approach requires three main components, similar to the basic optimizations formulation introduced in Chapter 3. These three components are displayed in Figure 5.1, and are explained as follows:

a. Design variables: In this case we seek to determine the joint profiles for the human skeleton.
b. Cost function: One or more cost functions are used to drive the behavior of the motion.
c. Constraints where any limitations on the motion or the interaction between the human and the environment are imposed. However, in this case we shall also scale the formulation introduced in Chapter 3 for posture prediction without dynamics to include the equations of motion...Newton's laws will now be acting on the entire motion to govern and subject any predictions to the laws of physics.

Figure 5.1 illustrates the three main components that formulate the predictive dynamics problem at an optimization formulation.

For the system, both the motion and the forces that cause the motion are unknown and treated as design variables in the optimization process. Equations of motion are treated as equality constraints instead of their direct numerical integration. The optimization problem is usually a large-scale nonlinear programming problem with many design variables and constraints. To solve the problem efficiently, modern optimization methods that take advantage of the structure of the problem are used.

The studied biosystem is represented as S and the corresponding mathematical model as M. The general equations of motion for the model M are written as:

$$f(\mathbf{q}, \dot{\mathbf{q}}, \ddot{\mathbf{q}}, t) = \tau \qquad (5.1)$$

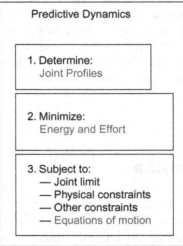

FIGURE 5.1

The three components involved in formulating the predictive dynamics problem.

where $\mathbf{q}, \dot{\mathbf{q}}, \ddot{\mathbf{q}} \in \mathbf{R}^N$ are the state variables, and $\tau \in \mathbf{R}^N$ are the generalized forces. This dynamics problem is defined over the time domain $\Omega = (T_0, T)$ with boundary $\Gamma = \{T_0, T\}, t \in \Omega$, t being the time and the symbol $(\cdot)$ indicating derivative with respect to t. The superscript N represents the number of DOF of model M.

Forward dynamics calculates the motion ($\mathbf{q}$, $\dot{\mathbf{q}}$ and $\ddot{\mathbf{q}}$) from the force τ by integrating Equation (5.1) with the specified initial conditions. In contrast, inverse dynamics computes the associated force τ that leads to a prescribed motion for the system. The two procedures are depicted in Figure 5.2. For simplicity, we use only $\mathbf{q}$ to represent kinematics of the system.

In practice, it is difficult to measure complete displacement ($\mathbf{q}$) and force (τ) histories accurately for a biosystem with many DOFs, especially involving a complex motion. This is because the experimental measurement is either not accurate enough or too expensive to achieve the required accuracy. However, the boundary conditions and some state response of the system might be available. In this case, neither forward dynamics nor inverse dynamics can be applied to the biosystem S directly. As a consequence, the predictive dynamics procedure is proposed to solve these types of problems. The basic idea is to formulate a nonlinear optimization problem based on the physics of motion (the dynamics of the motion). An appropriate performance measure (objective function) for the biosystem is defined and minimized subject to the available information about the system that imposes various constraints.

The ultimate intent of predictive dynamics is to enable the study of cause and effect on a human simulator, whereby a user defines the human body, body type, strength, and fatigue limits and is able to simulate an entire motion of the human. We illustrate the predictive dynamics problem using Figure 5.3 where the joint variables in this optimization problem are both displacement histories and force histories, and are both unknown; $\mathbf{g}$ are the constraints defined based on the available information Υ about the system S, such as physical constraints, collision, and

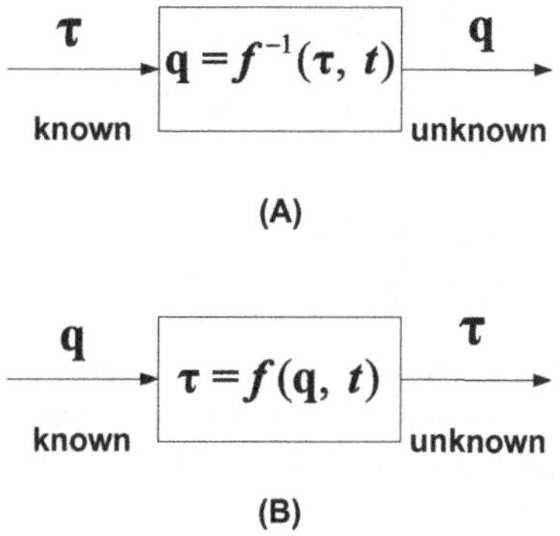

$$\tau \longrightarrow \boxed{\mathbf{q} = f^{-1}(\tau, t)} \longrightarrow \mathbf{q}$$

known   unknown

(A)

$$\mathbf{q} \longrightarrow \boxed{\tau = f(\mathbf{q}, t)} \longrightarrow \tau$$

known   unknown

(B)

FIGURE 5.2

Flowcharts of (A) forward dynamics and (B) inverse dynamics.

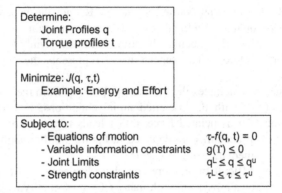

FIGURE 5.3

Illustration of the optimization problem in three main components.

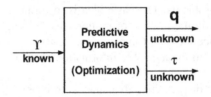

FIGURE 5.4

Flow chart of predictive dynamics.

others; $f(\mathbf{q}, \dot{\mathbf{q}}, \ddot{\mathbf{q}}, t) = \tau$ are the equations of motion; and $\mathbf{q}$ and τ are subject to their lower and upper bounds, respectively. This is called the predictive dynamics approach and is formulated for the system M as follows:

$$\min_{q, \tau} \ J(\mathbf{q}, \tau, \mathbf{t})$$
$$s.t.: \ \tau - f(\mathbf{q}, \dot{\mathbf{q}}, \ddot{\mathbf{q}}, \mathbf{t}) = 0$$
$$g(\Upsilon) \leq 0 \qquad\qquad (5.2)$$
$$\mathbf{q}^{L} \leq \mathbf{q} \leq \mathbf{q}^{U}$$
$$\tau^{L} \leq \tau \leq \tau^{U}$$

For the biosystem, the objective function is usually called the performance measure. The flowchart of predictive dynamics is shown in Figure 5.4.

Two challenging problems naturally arise for the predictive dynamics approach:

1. The functional form of the performance measure is unknown
2. The constraints are undetermined.

Before turning to these two important issues, the evaluation criterion for predictive dynamics is first proposed, i.e., how to validate the predictive dynamics solution of model M with that of the real system S.

Suppose q* and τ* represent the natural motion of the biosystem S. One way to quantify the accuracy of the predictive dynamics solution for the biosystem is to evaluate the percentage error of the residuals of the predicted force and displacement histories in a norm, as follows:

$$\varepsilon = \frac{\int_0^T (||\overline{\mathbf{q}}-\mathbf{q}^*|| + ||\overline{\tau}-\tau^*||)dt}{\int_0^T (||\mathbf{q}^*|| + ||\tau^*||)dt} \tag{5.3}$$

where $\overline{\mathbf{q}}$ and $\overline{\tau}$ are optimal values obtained from the predictive dynamics in Equation (5.2); T is the total time.

For human motion validation, some well-studied kinematics variables (angles and displacements) have been chosen as determinants to define a specific motion such as walking and running. For example, in walking motion the six determinants have been identified corresponding to the lower extremities and pelvic motion (Saunders et al., 1953). Therefore, instead of using all the joint angle and torque profiles in Equation (5.3), only the determinant and the corresponding torque profiles may be used to validate the task.

Furthermore, the predictive dynamics structure has very flexible optimization formulation in terms of constraints and performance measures for dynamic human motion prediction. For instance, the constraints allow one to model the boundary conditions and state response of the problem, and the performance measure allows one to study what drives human behavior. However, this assumes that the performance measure represents a physically significant quantity, not just a curve fit to predetermined data. Therefore, once the simulation model is validated, it can be used to show cause and effect, study an injury problem (reduce joint limit), analyze pathological motion, and so on.

5.3 Dynamic stability: zero-moment point

Dynamic stability of the human model is an important aspect of almost every task. Some tasks, such as rolling on the ground, may not require a stability criterion, but in most cases it is required.

Balance of the skeletal model is achieved by satisfying the zero-moment point (ZMP) constraint throughout the motion, if applicable. ZMP is briefly explained below and used in detail in the following chapter. In a balance state, ZMP coincides with the center of pressure (COP) where the resultant GRF acts (Xiang et al., 2010a,b).

The ZMP is a point on the ground at which the resultant tangential moments of the active forces are zero (Vukobratović and Borovac, 2004). It is used as the balance criterion for human walking. The forces on the system are divided into two categories: active forces and passive forces. *Active forces* include inertia, gravity, and applied external forces and moments. *Passive forces* are the ground reaction forces (GRF). The ZMP position and the GRF are calculated from the equations of motion by using a two-step algorithm: the resultant active forces and the ZMP location are calculated in the first step, and the GRF are calculated in the second step. The basic idea of this algorithm is to obtain the GRF from the resultant active forces by imposing the overall equilibrium of the digital human model at ZMP. This algorithm was first introduced in Xiang et al. (2007) to simulate 3D human gait, and details were presented elsewhere (Xiang, 2008; Xiang et al., 2009b). It is outlined below but will be expanded upon in the next chapter.

The following calculations are performed at each time instant:

Step 1: Calculation of resultant active forces and ZMP

1.1. Given q, $\dot{q}$, $\ddot{q}$ for each DOF, the global resultant active forces ($\mathbf{M}^o$, $\mathbf{F}^o$) at the origin in the global coordinate system (o-xyz), excluding GRF, are calculated from equations of motion using inverse dynamics.

1.2. The ZMP position is calculated from its definition using the global resultant active forces and assuming the feet to be on the level ground, as follows:

$$y_{zmp} = 0; \quad x_{zmp} = \frac{M_z^o}{F_y^o}; \quad z_{zmp} = \frac{-M_x^o}{F_y^o} \tag{5.4}$$

where $\mathbf{M}^o = \begin{bmatrix} M_x^o & M_y^o & M_z^o \end{bmatrix}^T$ and $\mathbf{F}^o = \begin{bmatrix} F_x^o & F_y^o & F_z^o \end{bmatrix}^T$.

1.3. After obtaining the ZMP position, the resultant active forces at ZMP ($\mathbf{M}^{zmp}$, $\mathbf{F}^{zmp}$) are computed using the equilibrium conditions as follows:

$$\mathbf{M}^{zmp} = \mathbf{M}^o + {}^o\mathbf{r}_{zmp} \times \mathbf{F}^o; \quad \mathbf{F}^{zmp} = \mathbf{F}^o \tag{5.5}$$

where ${}^o\mathbf{r}_{zmp}$ is the ZMP position in the global coordinate system obtained from Equation (5.4).

Step 2: Calculation of GRF

The value and location of GRF are calculated from the equilibrium between the resultant active forces and passive forces at the ZMP:

$$\mathbf{M}^{GRF} + \mathbf{M}^{zmp} = 0; \quad \mathbf{F}^{GRF} + \mathbf{F}^{zmp} = 0; \quad {}^o\mathbf{r}_{GRF} - {}^o\mathbf{r}_{zmp} = 0 \tag{5.6}$$

All active forces (gravity, inertia and applied external forces) and passive forces (GRF) are applied to the entire human body model, and the equations of motion are used to obtain the real joint torques. In this case, the global forces are zero and the equilibrium of the system is satisfied automatically.

There are two prominent features to calculate GRF in the foregoing two-step algorithm:

1. GRF are explicitly calculated from joint kinematics; and differentiation of GRF with respect to the system kinematics is easily achieved.

2. Since GRF are directly obtained from the system equilibrium conditions, the equilibrium is satisfied while evaluating GRF using equations of motion (Xiang et al., 2007). This two-step algorithm is called *GRF equilibrium inclusion* formulation.

5.4 Performance measures

In Chapter 3, we have presented some basic cost functions used to predict posture. We remind the reader that these human performance measures were used as cost functions to drive the behavior of the system, which means the behavior of the human's motion. However, these cost functions in Chapter 3 did not take into consideration any dynamic or inertial quantities. They simply did not take the dynamics of the motion into effect. In practice, the performance measures include many different kinematics and dynamics criteria such as time minimization, torque optimization, energy minimization, jerk minimization, and others.

In this chapter, we add a few more that are far more effective in the dynamics simulation for human motion.

1. Dynamic effort

$$f = \int_0^T \tau \cdot \tau \, dt \qquad (5.7)$$

which is defined as the integration of squares of all joint torques over time. The value of this integration is a torque square summation, which is very closely related to the energy of the system.

2. Mechanical energy

$$f = \int_0^T |\tau \cdot \dot{q}| dt \qquad (5.8)$$

which measures the mechanical energy for the mechanical system.

3. Metabolic energy

$$f = \int_0^T \dot{E} dt \qquad (5.9)$$

where $\dot{E}$ is the rate of total metabolic energy.

4. Jerk

$$f = \int_0^T \dot{\tau} \cdot \dot{\tau} \, dt \qquad (5.10)$$

which is defined as the integration of squares of all joint torque derivatives. An alternative form of jerk is to evaluate the derivative of acceleration instead of joint torque. Minimization of jerk gives a smoother motion.

5. Stability

$$f = \int_0^T S \, dt \tag{5.11}$$

where S represents the stability quantity, which can be defined in different ways. One definition is the deviation of ZMP position from the center of the support polygon or a prescribed ZMP trajectory (Huang et al., 2001; Xiang et al., 2010b). The second definition is the deviation of the trunk from vertical position (Gubina et al., 1974; Kim et al., 2008).

6. Maximum absolute value of joint torque

$$f = \max_i \{|\tau_i|\} \tag{5.12}$$

A common approach for treating cost function in Equation (5.12) is to introduce an additional unknown parameter λ (Rasmussen et al., 2001; Xiang et al., 2009a):

$$
\begin{aligned}
Min. \quad & \lambda \\
s.t. \quad & \tau_i \leq \lambda, \\
& -\tau_i \leq \lambda, \quad i = 1, 2, \ldots, n
\end{aligned}
\tag{5.13}
$$

7. Dynamic effort as a cost function

A performance measure that is well studied in the biomechanics literature is the minimizing of the squares of all actuating torques or minimizing the maximum torque for all joints.

The *dynamic effort*, which is represented as time integral of the squares of all joint torques, is used as a performance measure to be minimized for the walking motion. This is also sometimes called the total torque effort. The predicted motion depends strongly on the adopted objective function F. As an example, dynamic effort sometime used as the performance criterion for the walking problem can be written as:

$$F(q) = \int_{t=0}^T \left(\frac{\tau(q, t)}{|\tau|_{max}} \right)^T \cdot \left(\frac{\tau(q, t)}{|\tau|_{max}} \right) dt \tag{5.14}$$

where $|\tau|_{max}$ is the maximum absolute value of joint torque limit.

5.5 Inner optimization

An alternative way to define the performance measure is to use the inner optimization (nested optimization) method. The basic idea is to construct a local search space of cost functions S_J with a specific functional form based on some insight into the physical processes governing the biosystem.

$$S_J(q, \tau, t|p^*) \approx J(q, \tau, t) \tag{5.15}$$

where p^* is the parameter vector that needs to be determined.

For instance, a performance measure for the biosystem has been identified as the sum of squares of all joint torques with coefficients **p** as in Equation (5.16).

$$S_J(q, \tau, t|p) = \int_0^T \sum_{i=1}^{n_\tau} p_i \tau_i^2 \, dt \tag{5.16}$$

where

$$\sum_{i=1}^{n_\tau} p_i = 1 \quad and \quad p_i \geq 0 \tag{5.17}$$

The parameters p are determined by solving the inner optimization problem as defined in Equation (5.18) so that the exact performance measure can be identified.

$$\begin{aligned} &\min_{p} \quad \varepsilon \\ &s.t.: \quad \mathbf{h(p)} \leq \mathbf{0} \\ &\qquad \min_{q, \tau} \quad S_J(\mathbf{q}, \tau, \mathbf{t|p}) \\ &\qquad s.t.: \quad \tau - f(\mathbf{q}, \dot{\mathbf{q}}, \ddot{\mathbf{q}}, \mathbf{t}) = \mathbf{0} \\ &\qquad \mathbf{g(\Upsilon)} \leq \mathbf{0} \\ &\qquad \mathbf{q}^L \leq \mathbf{q} \leq \mathbf{q}^U \\ &\qquad \tau^L \leq \tau \leq \tau^U \end{aligned} \tag{5.18}$$

where h(p) $\leq$ **0** are the possible equality or inequality constraints on the parameters p satisfying the normalization and non-negativity conditions. ε is the error defined in Equation (5.3). The process of identifying the unknown performance measure is transformed to find the parameters p that will minimize the error ε.

5.6 Constraints

Constraints are formulated based on the available information Υ about the biosystem. In general, two types of constraints are included in this set: (i) boundary conditions, and (ii) state response at some time points, $q(t_j), t_j \in \Omega$, obtained from either experiments or observations. In addition, boundary conditions consist of time boundary, $q_{t_j}, t_j \in \Gamma$, and geometrical boundary, $X(q_{t_j}), t_j \in \Omega \cup \Gamma$, where **X** represents the global Cartesian coordinates that capture the geometrical environment for the biosystem. For example, given initial and final postures, a walking task is performed to predict the walking motion between the two postures. The initial and final postures are the time boundaries, and the ground is formulated as a geometrical boundary. However, there are many options for state response constraints based on available information about the walking motion, such as transition posture between single support and double support phases, knee flexion

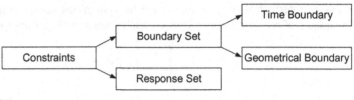

FIGURE 5.5

Constraints for a biosystem.

angle at mid-swing time point, etc. The overall set of constraints is depicted in Figure 5.5.

5.6.1 Feasible set

The feasible set of solutions for the problem is an important issue for predictive dynamics. An infeasible set will result in a null solution space for the system. This situation should always be avoided while formulating a predictive dynamics problem. For a biosystem, feasibility of all the constraints can be tested by solving the predictive dynamics problem with a constant objective function as follows:

$$
\begin{aligned}
\min_{q,\,\tau} \;\; & J(\mathbf{q},\,\tau,\,\mathbf{t}) \equiv c \\
s.t.: \;\; & \tau - f(\mathbf{q},\,\dot{\mathbf{q}},\,\ddot{\mathbf{q}},\,\mathbf{t}) = \mathbf{0} \\
& \mathbf{g}(\Upsilon) \leq \mathbf{0} \\
& \mathbf{q}^L \leq \mathbf{q} \leq \mathbf{q}^U \\
& \tau^L \leq \tau \leq \tau^U
\end{aligned}
\tag{5.19}
$$

where c is a constant.

The solution of Equation (5.19) implies that the output set $(q^f,\,\tau^f)$ satisfies all linear and nonlinear constraints, but does not optimize any performance measure for the biosystem. This is a feasible solution of the predictive dynamics problem. There are two purposes of obtaining a feasible solution for the system: one is to test the feasibility of all the constraints, and the other is to get a solution that might be used as a good initial guess for the predictive dynamics with a physical performance measure.

5.6.2 Minimal set of constraints

It is obvious that the more information about the biosystem that is available, the more accurate the predictive dynamics solution is. As an extreme case, all the displacement and force histories can be available in the time domain, $\Omega \cup \Gamma$. However, in most cases only minimal information about the biosystem is available so that predictive dynamics seeks the minimal constraint set $\mathbf{g}(\Upsilon_{minimal})$ and an appropriate performance measure to simulate the applied force and response histories for the biosystem, as follows:

$$\min_{q, \tau} \quad J(\mathbf{q}, \tau, \mathbf{t})$$

$$s.t.: \quad \tau - f(\mathbf{q}, \dot{\mathbf{q}}, \ddot{\mathbf{q}}, \mathbf{t}) = \mathbf{0}$$

$$\mathbf{g}(\Upsilon_{minimal}) \leq \mathbf{0} \qquad (5.20)$$

$$\mathbf{q}^L \leq \mathbf{q} \leq \mathbf{q}^U$$

$$\tau^L \leq \tau \leq \tau^U$$

The minimal constraint set depends on the complexity of the biosystem and the motion to be simulated. For a simple motion, boundary conditions alone might be enough to reveal the entire motion; in this case, the minimal constraint set includes only boundary conditions. In contrast, for a complex motion some state responses between the boundaries need to be known to simulate the real motion. Therefore, these state responses have to be included in the minimal constraint set.

5.7 Types of constraints

5.7.1 Time-dependent constraints

5.7.1.1 Joint limits

To avoid hyperextension, the joint limits are taken into account in the formulation. The joint limits representing the physical range of motion are:

$$\mathbf{q}^L \leq \mathbf{q}(t) \leq \mathbf{q}^U, \quad 0 \leq t \leq T \qquad (5.21)$$

where $\mathbf{q}^L$ are the lower joint limits and $\mathbf{q}^U$ the upper limits. Limits on major joints are presented in Table 5.1.

Joint limit constraint is also used to "freeze" a DOF by setting its lower bound and upper bound to the neutral angle (the natural angle at rest) instead of eliminating this DOF from the skeleton model. Changing lower or upper joint limit constraints in conjunction with strength constraints at a single joint can simulate a disability and will cause the model to respond differently.

5.7.1.2 Torque limits

Each joint torque is also bounded by its physical limits (strength), which are obtained from several references (Cahalan et al., 1989; Gill et al., 2002; Kaminski et al., 1999; Kumar, 1996):

$$\tau^L \leq \tau(t) \leq \tau^U, \quad 0 \leq t \leq T \qquad (5.22)$$

where τ^L are the lower torque limits and τ^U the upper limits.

Table 5.1 Major Joint Angle Limits[a,b]

Joints	Joint Angle Limits (Degree)	
	Lower Limit	Upper Limit
Ankle (dorsi/plantar)	−20	54.5
Knee (extension/flexion)	7	138
Hip (flexion/extension)	−102	41
Hip (abduct/adduct)	−46	34
Hip (external/internal)	−49	32
Spine (tilt)	−11	11
Spine (bend)	−9.5	21
Spine (rotate)	−13.5	13.5
Shoulder (aft/fore)	−19	111
Shoulder (adduct/abduct)	−23	123.5

[a]Zero joint angles correspond to home configuration as depicted in Figure 7.2.
[b]Joint coupling motions are not considered.

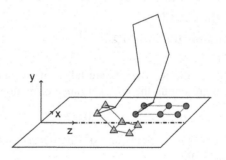

FIGURE 5.6

Foot ground penetration conditions.

5.7.1.3 Ground penetration

Walking is characterized with unilateral contact between the foot and ground as shown in Figure 5.6. While the foot contacts the ground, the height and velocity of contacting points (circles) are zero. In contrast, the height of other points (triangles) on the foot is greater than zero.

Therefore, the ground penetration constraints are formulated as follows:

$$y_i(t) = 0, \quad \dot{x}_i(t) = 0, \quad \dot{y}_i(t) = 0, \quad \dot{z}_i(t) = 0, \quad i \in \Omega$$
$$y_i(t) > 0, \qquad i \notin \Omega, \qquad 0 \leq t \leq T \tag{5.23}$$

where Ω is the set of contacting points as illustrated in Figure 5.6.

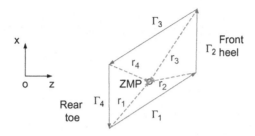

FIGURE 5.7

Foot support polygon (top view).

5.7.1.4 Dynamic balance

The dynamic balance is achieved by enforcing the ZMP to remain within the foot support polygon (FSP) as depicted in Figure 5.7, where Γ is a vector along the boundary of FSP and r is the position vector from a vertex of the FSP to ZMP.

The ZMP constraint is mathematical expressed as follows:

$$(\mathrm{r}_i \times \Gamma_i) \cdot \mathrm{n_y} \leq 0, \quad i = 1, \ldots, 4 \tag{5.24}$$

where $\mathrm{n_y}$ is the unit vector along the y axis.

5.7.1.5 Self-avoidance

Self-avoidance is considered in the current formulation to prevent penetration of the arm in the body. A sphere-filling algorithm is used to formulate this constraint as shown in Figure 5.8, as follows:

$$d(\mathrm{q}, t) - r_1 - r_2 \geq 0, \quad 0 \leq t \leq T \tag{5.25}$$

where r_1 is a constant radius to represent the wrist, and r_2 is another radius to represent the hip; d is the distance between wrist and hip.

5.7.2 Time-independent constraints

5.7.2.1 Symmetry conditions

The gait simulation starts from the left heel strike and ends with the right heel strike. The initial and final postures and velocities should satisfy the symmetry conditions to generate continuous walking motion. These conditions are expressed as follows:

$$
\begin{aligned}
q_L(0) - q_R(T) &= 0 & \dot{q}_L(0) - \dot{q}_R(T) &= 0 \\
q_{Sx}(0) - q_{Sx}(T) &= 0 & \dot{q}_{Sx}(0) - \dot{q}_{Sx}(T) &= 0 \\
q_{Sy}(0) + q_{Sy}(T) &= 0 & \dot{q}_{Sy}(0) + \dot{q}_{Sy}(T) &= 0 \\
q_{Sz}(0) + q_{Sz}(T) &= 0 & \dot{q}_{Sz}(0) + \dot{q}_{Sz}(T) &= 0
\end{aligned}
\tag{5.26}
$$

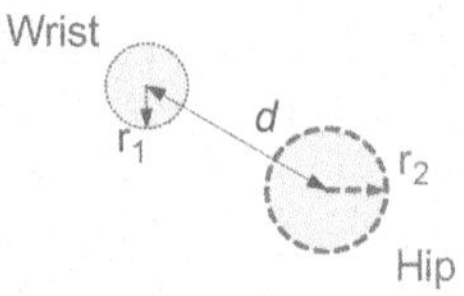

FIGURE 5.8

Self-avoidance constraint between the wrist and hip.

where subscripts L and R represent the DOFs of the leg, arm and shoulder joints which satisfy the symmetry conditions with the contra-lateral leg, arm and shoulder joints; the subscript S represents the DOFs of spine, neck and global joints which satisfy their symmetry conditions at the initial and final times; x, y, z are the global axes.

5.7.2.2 *Ground clearance*

To avoid foot drag motion, ground clearance constraint is imposed during the walking motion. Instead of controlling the maximum height of the swing leg, the maximum knee flexion at mid-swing is used to formulate ground clearance constraint. Biomechanical experiments have shown that the maximum knee flexion of normal gait is around 60 degrees regardless of the subject's age and gender. This constraint is expressed as

$$-\varepsilon \leq q_{knee} - 60 \leq \varepsilon, \qquad t = t_{midswing} \tag{5.27}$$

where ε is a small range of motion, i.e., $\varepsilon = 5$ degrees.

5.8 Discretization and scaling

The predictive dynamics problem in Equation (5.2) is actually an optimal control problem with boundary conditions and some state constraints. The classical method to solve the optimal control problem is to derive the optimality condition for the continuous variable optimization problem. However, beyond boundary conditions the continuous method has difficulty dealing with discrete state constraints. The most efficient way to solve a complex optimal control problem is to use nonlinear optimization techniques. The basic idea is to discretize the governing equations of motion using a suitable numerical method and define finite dimensional approximation for the state and control variables. This process transforms the system differential equations into algebraic equations with parametric representation of the state and control variables. The performance measures and the constraints are also evaluated in terms of discrete state and control values. Therefore, the original optimal control problem is transformed into a nonlinear programming (NLP) problem.

The time domain is first discretized into n intervals with step size h^i, as follows:

$$0 = t_0 \leq t_1 \leq \dots t_{n-1} \leq t_n = T \quad and \quad h^i = t_{i+1} - t_i \tag{5.28}$$

The discretized state q_h and force τ_h can be expressed in terms of interpolating functions and discrete nodal DOF (control points) P_q and P_τ.

$$q_h = q(P_q, h); \quad \tau_h = \tau(P_\tau, h) \tag{5.29}$$

Thus, the discretized predictive dynamics problem is formulated as:

$$\min_{q,\,\tau} \quad J(\mathbf{q}_h,\, \tau_h,\, \mathbf{t}_h)$$
$$s.t.: \quad \tau_h - f(\mathbf{q}_h, \dot{\mathbf{q}}_h, \ddot{\mathbf{q}}_h, \mathbf{t}_h) = 0$$
$$\mathbf{g}(\Upsilon_h) \leq 0 \tag{5.30}$$
$$\mathbf{q}^L \leq \mathbf{q}_h \leq \mathbf{q}^U$$
$$\tau^L \leq \tau_h \leq \tau^U$$

In general, all the unknowns and the equations of motion should be scaled to improve the numerical performance of the nonlinear optimization solver. Appropriate scale factors are chosen so as to obtain quantities that have the same magnitude order, $O(1)$. It is noted that scaling of a constraint does not change the constraint boundary, so it has no effect on the optimum solution.

$$\min_{q,\,\tau} \quad J(s^q \mathbf{q}_h,\, s^\tau \tau_h,\, s^t \mathbf{t}_h)$$
$$s.t.: \quad s^\tau \tau_h - f(s^q \mathbf{q}_h, s^q \dot{\mathbf{q}}_h, s^q \ddot{\mathbf{q}}_h, s^t \mathbf{t}_h) = 0$$
$$\mathbf{g}(\Upsilon_h) \leq 0 \tag{5.31}$$
$$s^q \mathbf{q}^L \leq s^q \mathbf{q}_h \leq s^q \mathbf{q}^U$$
$$s^\tau \tau^L \leq s^\tau \tau_h \leq s^\tau \tau^U$$

5.9 Numerical example: single pendulum

Before applying predictive dynamics to a large problem, such as a human with many DOF, we will examine the method's implementation in a simpler and more basic problem to illustrate its power. The example below considers a simple pendulum.

5.9.1 Description of the problem

The natural swinging motion of a single pendulum subjected to external torque is considered. The swinging motion is first treated as a forward dynamics problem with the known external force and solved by the multi-body dynamics solver ADAMS. The solution is assumed to be the true response of the system used to evaluate the results obtained with the predictive dynamics formulation. Predictive dynamics is implemented based on the available information about the system. Four cases are examined with predictive dynamics as listed in Table 5.2.

Table 5.2 Four Cases of Swing Motion Examined with Predictive Dynamics

Motion	Case	Available Information
Simple swing without oscillation	1	Boundary conditions
Oscillating motion	2	Boundary conditions
	3	Boundary conditions and response at one point
	4	Boundary conditions and response at two points

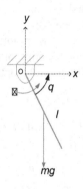

FIGURE 5.9

Single pendulum.

The pendulum pivots at the point O as shown in Figure 5.9. The equation of motion for a rigid bar subject to external torque is given as

$$I\ddot{q} + mg\frac{l}{2}\cos q = \tau \qquad (5.32)$$

where I is the moment of inertia, m is the mass, l is the length, q is the joint angle, and τ is the external torque. The external torque is assumed to be a sinusoidal function given as

$$\tau = 0.1\sin 5t \quad (\text{Nm}) \qquad (5.33)$$

The geometrical and physical parameters of the rigid bar are taken as $I = 0.0267 \text{ kgm}^2$, $m = 0.5 \text{ kg}$, and $l = 0.4 \text{ m}$. With the initial condition $q(0) = 0$ and $\dot{q}(0) = 0$, the forward dynamics is solved by the ADAMS Runge-Kutta solver, as shown in Figure 5.10, which is considered the true solution of the system.

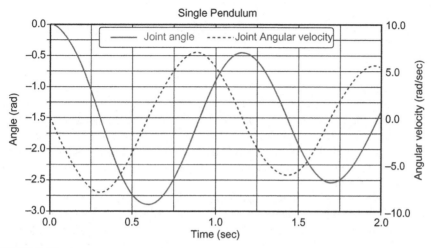

Single Pendulum

FIGURE 5.10

Forward dynamics solved by ADAMS.

5.9.2 Simple swing motion with boundary conditions—PD solution

The swinging motion without oscillation is studied with the initial and final conditions. The total time is randomly selected as $T = 0.43$ s (less than first half period of oscillation), and the single pendulum starts at rest in the horizontal position and ends up with the final conditions that are obtained from Figure 5.10 as $q(T) = -2.40$ rad and $\dot{q}(T) = -5.86$ rad/s. The swinging motion is driven by the external torque and gravity. Besides boundary condition and total travel time T, neither external torque nor joint angle is known. Therefore, the predictive dynamics problem is formulated as in Equation (5.31) to reveal the natural swing motion of the single pendulum.

$$\text{Minimize}\quad J(q, \tau, t)$$

$$\text{Subject to}\quad I\ddot{q} + mg\frac{l}{2}\cos q = \tau$$

$$q(0) = 0, \dot{q}(0) = 0 \tag{5.34}$$
$$q(T) = -2.40, \dot{q}(T) = -5.86, T = 0.43$$
$$-\pi \leq q \leq \pi$$
$$-10 \leq \tau \leq 10$$

Treating q and τ as design variables, four performance measures are tested as follows

$$J_1(q, \tau, t) = \int_{t=0}^{T} \tau \cdot \tau dt \tag{5.35}$$

$$J_2(q, \tau, t) = \max_{t \in [0,T]} \tau \qquad (5.36)$$

$$J_3(q, \tau, t) = T \qquad (5.37)$$

$$J_4(q, \tau, t) = c \qquad (5.38)$$

The first performance measure is to minimize the integral of squares of the joint torque for the entire time domain, which is a form of mechanical energy. The second is to minimize the maximum torque over the entire time domain, and the third is to minimize the total travel time, T, subjected to the same boundary conditions. The final performance measure is to solve for only a feasible solution where c is a constant.

The optimization problem is discretized into a nonlinear programming problem and then solved by SNOPT with various performance measures as defined in Equations (5.35−5.38). The optimal solution yields the joint angle, velocity, and external torque history as depicted in Figures 5.11, 5.12, and 5.13, respectively.

Figures 5.12 and 5.13 show that the performance measures torque-square and min−max torque successfully predict joint angle and velocity response. However, only torque-square predicts joint torque correctly. Minimizing the total time or a constant fails to predict the response of the dynamic system. This is explained by the fact that the natural motion always obeys an energy-saving rule so the energy-related performance measure is more appropriate for predicting the dynamic motion.

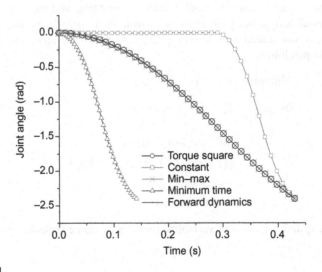

FIGURE 5.11

Joint angle prediction of the single pendulum, Case 1.

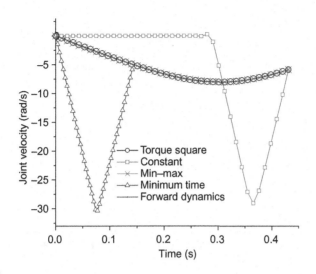

FIGURE 5.12

Joint velocity prediction of the single pendulum, Case 1.

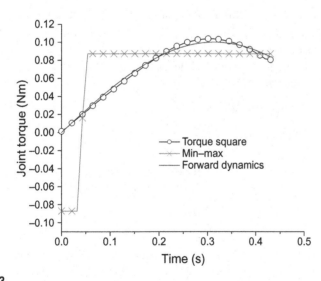

FIGURE 5.13

Joint torque prediction of the single pendulum, Case 1.

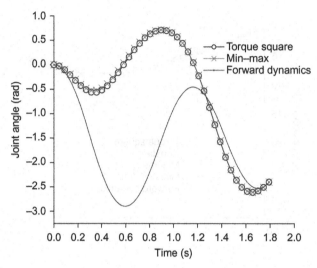

FIGURE 5.14

Joint angle prediction of the single pendulum, Case 2.

5.9.3 Oscillating motion with boundary conditions—PD solution

Oscillating pendulum makes the motion more complex. The predictive dynamics approach is examined in this case by extending the final time to $T = 1.79$ s (more than one and one-half period). The optimization formulation is similar to Equation (5.33) except for the final conditions.

$$\text{Minimize } J(q, \tau, t)$$

$$\text{Subject to } I\ddot{q} + mg\frac{l}{2}\cos q = \tau$$

$$q(0) = 0, \dot{q}(0) = 0 \tag{5.39}$$
$$q(T) = -2.40, \dot{q}(T) = -2.85, T = 1.79$$
$$-\pi \leq q \leq \pi$$
$$-10 \leq \tau \leq 10$$

Note from the results of the previous section that the performance measure of minimum total time or a constant is not appropriate for predicting the natural swinging motion of the single pendulum. Thus, only torque squares and min−max formulations are tested as performance measures for the present case. The optimized joint angle, velocity, and applied torque are given in Figures 5.14−5.16.

For the oscillating motion, the predictive dynamics fails to predict joint angle, velocity, and torque histories with only the boundary conditions specified. Although an energy-related performance measure is chosen, predictive dynamics

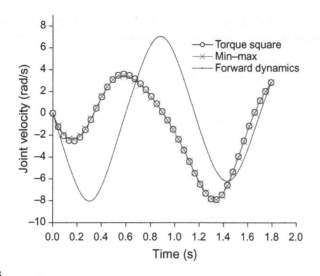

FIGURE 5.15

Joint velocity prediction of the single pendulum, Case 2.

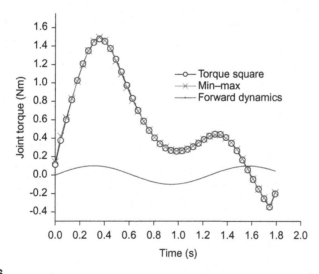

FIGURE 5.16

Joint torque prediction of the single pendulum, Case 2.

cannot predict the true response due to lack of necessary information (constraints) on the dynamic system between the boundaries. This is because both force and motion are unknowns in the optimization formulation. Although the boundary conditions are satisfied, it may predict different motion between the boundaries.

5.9.4 Oscillating motion with boundary conditions and one state-response constraint—PD solution

Adding one state constraint, $q(0.76) = -2.44\,\text{rad}$(obtained from Figure 5.16), to the optimization formulation in the previous section, predictive dynamics is formulated as:

$$\text{Minimize} \quad J(q, \tau, t)$$

$$\text{Subject to} \quad I\ddot{q} + mg\frac{l}{2}\cos q = \tau$$

$$q(0) = 0, \dot{q}(0) = 0$$
$$q(0.76) = -2.44 \tag{5.40}$$
$$q(T) = -2.40, \dot{q}(T) = -2.85, T = 1.79$$
$$-\pi \leq q \leq \pi$$
$$-10 \leq \tau \leq 10$$

Solving the above optimization problem with the two performance measures, the corresponding joint angle, velocity, and torque are obtained, as shown in Figures 5.17−5.19.

In Figures 5.16−5.18, min−max performance closely predicts joint angle and velocity histories of the system. However, it has a bang-bang type prediction for the joint torque. Torque-square performance only predicts the trends of joint angle and velocity and fails to predict joint torque.

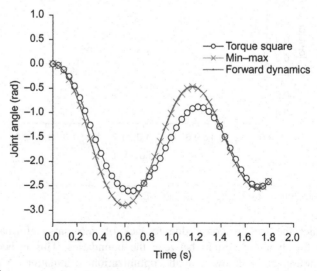

FIGURE 5.17

Joint angle prediction of the single pendulum, Case 3.

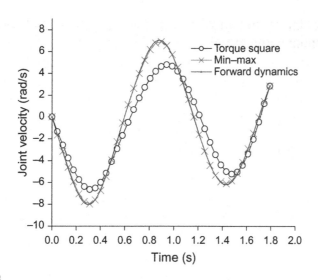

FIGURE 5.18

Joint velocity prediction of the single pendulum, Case 3.

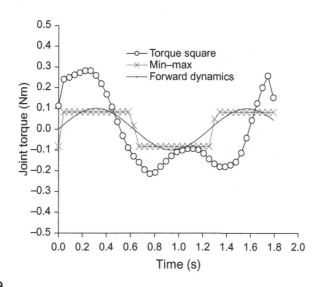

FIGURE 5.19

Joint torque prediction of the single pendulum, Case 3.

5.9.5 Oscillating motion with boundary conditions and two state-response constraints

Besides boundary conditions, two more state responses, $q(0.76) = -2.44$ rad and $q(1.21) = -0.493$ rad, are imposed as additional constraints for the optimization problem. The predictive dynamics problem is defined as:

$$\text{Minimize} \quad J(q, \tau, t)$$

$$\text{Subject to} \quad I\ddot{q} + mg\frac{l}{2}\cos q = \tau$$

$$
\begin{aligned}
&q(0) = 0, \dot{q}(0) = 0 \\
&q(0.76) = -2.44 \\
&q(1.21) = -0.493 \\
&q(T) = -2.40, \dot{q}(T) = -2.85, T = 1.79 \\
&-\pi \le q \le \pi \\
&-10 \le \tau \le 10
\end{aligned}
\qquad (5.41)
$$

Predicted joint angle, velocity, and torque are given in Figures 5.20–5.22.

With two more state-response constraints, the predictive dynamics closely reveals the joint angle, velocity, and torque histories. It is important to note that the min−max performance measure still has a bang-bang type of joint torque. Both joint angle and velocity are identified by torque-square and min−max performance measures.

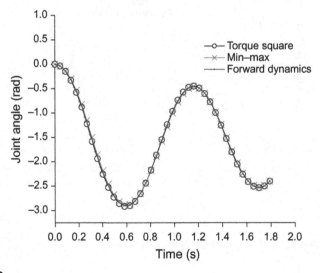

FIGURE 5.20

Joint angle prediction of the single pendulum, Case 4.

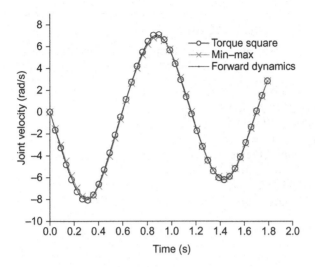

FIGURE 5.21

Joint velocity prediction of the single pendulum, Case 4.

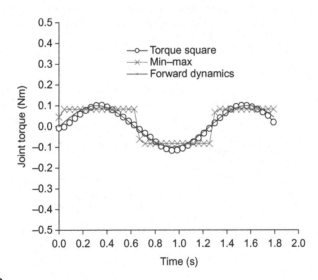

FIGURE 5.22

Joint torque prediction of the single pendulum, Case 4.

As one can see, predictive dynamics yields numerical results that are consistent with those obtained from a forward dynamics computational method.

A more detailed treatment of two specific tasks is presented in Chapter 7 for the prediction of the biomechanics of walking and in Chapter 8 for the prediction of the biomechanics of lifting.

5.10 **Example formulations**

The next section presents summary formulations for various tasks of predictive dynamics. The major components for each task are presented in a figure:

a. Walking [Figure 5.23A]
b. Box lifting [Figure 5.23B]
c. Stairs climbing [Figure 5.23C]
d. Throwing [Figure 5.23D]
e. Ladder climbing [Figure 5.23E]
f. Running [Figure 5.23F]
g. Kneeling [Figure 5.23G]
h. Jumping [Figure 5.23H]

5.11 **Concluding remarks**

Predictive dynamics is a new method for simulating the response of a system given a certain action. The most significant issue that summarizes this chapter is characterized by this method's ability to solve for the motion without the need to integrate the equations of motion. Unlike typical forward dynamics methods

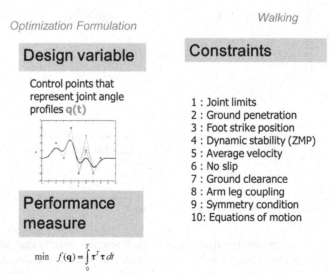

Optimization Formulation   *Walking*

Design variable

Control points that represent joint angle profiles q(t)

Constraints

1 : Joint limits
2 : Ground penetration
3 : Foot strike position
4 : Dynamic stability (ZMP)
5 : Average velocity
6 : No slip
7 : Ground clearance
8 : Arm leg coupling
9 : Symmetry condition
10: Equations of motion

Performance measure

$$\min \quad f(\mathbf{q}) = \int_0^T \tau^T \tau \, dt$$

FIGURE 5.23A

Walking.

Optimization Formulation

Box Lifting

Design variable

Control points that
represent joint angle
profiles q(t)

Performance measure

$$\min \quad f(\mathbf{q}) = \int_0^T \boldsymbol{\tau}^T \boldsymbol{\tau} \, dt$$

Constraints

Time-dependent

1 : Joint limits
2 : Ground penetration
3 : Foot standing position
4 : Dynamic stability (ZMP)
5 : Collision avoidance
6 : Torque limits
7 : Hand distance
8 : Hand orientation (1) hands parallel
9 : Hand orientation (2) hand normal to box
10: Vision

Time-independent

1 : Initial hand positions
2 : Final hand positions
3 : Initial static condition
4 : Final static condition

FIGURE 5.23B

Box lifting.

Optimization Formulation

Stairs Climbing

Design variables

Joint angle profiles
q(t)

Performance measure

$$\min \quad f(\mathbf{q}) = \int_0^T \boldsymbol{\tau}^T \boldsymbol{\tau} \, dt$$

Constraints

General

1 : Joint angle limits
2 : Torque limits
3 : Equations of motion

Task-specific

1 : Dynamic stability (ZMP)
2 : No slip
3 : Foot strike position
4 : Arm leg coupling
5 : Self avoidance
6 : Ground penetration
7 : Symmetry/Continuity condition
8 : Ground clearance
9 : Stairs avoidance constraint

FIGURE 5.23C

Stairs climbing.

Optimization Formulation

Throwing

Design variables

Joint angle profiles
q(t)

Object flight time
T

Performance measure

$$\min \quad f(\mathbf{q}) = \int_0^T \boldsymbol{\tau}^T \boldsymbol{\tau} \, dt$$

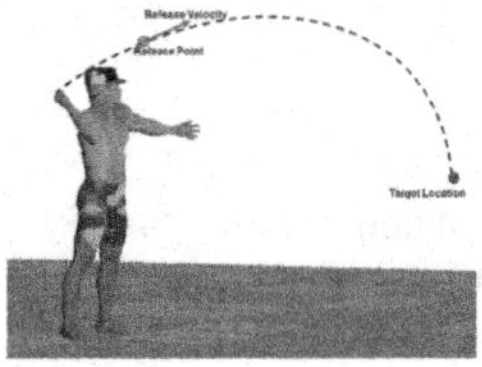

Constraints

General

1 : Joint angle limits
2 : Torque limits
3 : Equations of motion

Task-specific

1 : Dynamic stability (ZMP)
2 : Ground penetration
3 : Foot strike position
4 : Target within visual field
5 : Overhand throw (positive y-velocity)
6 : Monotonic hand path (positive z-velocity)
7 : Static initial condition
8 : Initial posture
9 : Projectile equation
10: Hand release orientation

FIGURE 5.23D

Throwing.

Optimization Formulation

Ladder Climbing – phase 1

Design variables

Joint angle profiles
q(t)

Reaction forces (for phase 2)
R(t)

Performance measure

$$\min \quad f(\mathbf{q}) = \int_0^T \boldsymbol{\tau}^T \boldsymbol{\tau} \, dt$$

Constraints

General

1 : Joint angle limits
2 : Torque limits
3 : Equations of motion

Task-specific

1 : Hands and feet contact positions
2 : Periodic joint conditions
3 : Rung-body penetration avoidance
4 : ZMP (for phase 2)
5 : Friction limit (for phase 2)

FIGURE 5.23E

Ladder climbing.

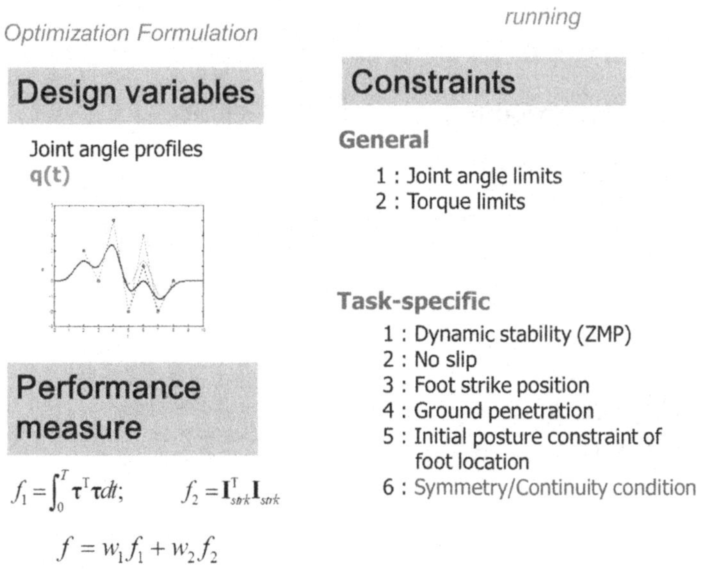

Optimization Formulation

running

Design variables

Joint angle profiles
q(t)

Performance measure

$$f_1 = \int_0^T \boldsymbol{\tau}^T \boldsymbol{\tau} dt; \qquad f_2 = \mathbf{I}_{strk}^T \mathbf{I}_{strk}$$

$$f = w_1 f_1 + w_2 f_2$$

Constraints

General

1 : Joint angle limits
2 : Torque limits

Task-specific

1 : Dynamic stability (ZMP)
2 : No slip
3 : Foot strike position
4 : Ground penetration
5 : Initial posture constraint of foot location
6 : Symmetry/Continuity condition

FIGURE 5.23F

Running.

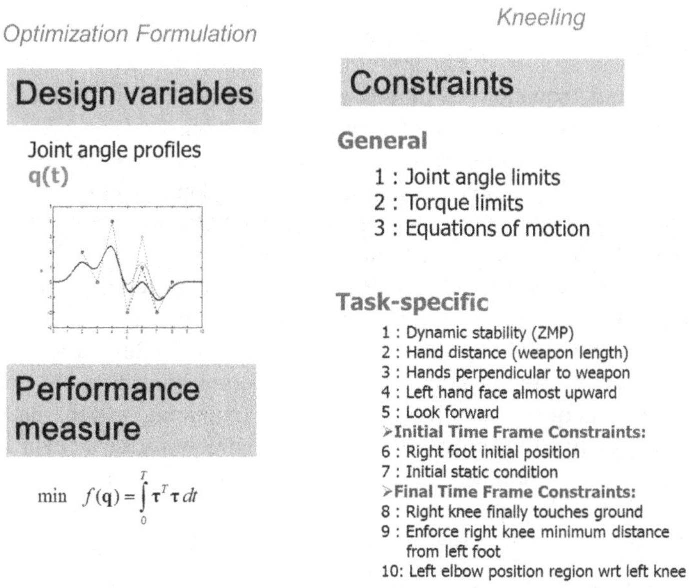

Optimization Formulation

Kneeling

Design variables

Joint angle profiles
q(t)

Performance measure

$$\min \quad f(\mathbf{q}) = \int_0^T \boldsymbol{\tau}^T \boldsymbol{\tau} \, dt$$

Constraints

General

1 : Joint angle limits
2 : Torque limits
3 : Equations of motion

Task-specific

1 : Dynamic stability (ZMP)
2 : Hand distance (weapon length)
3 : Hands perpendicular to weapon
4 : Left hand face almost upward
5 : Look forward
➢**Initial Time Frame Constraints:**
6 : Right foot initial position
7 : Initial static condition
➢**Final Time Frame Constraints:**
8 : Right knee finally touches ground
9 : Enforce right knee minimum distance from left foot
10: Left elbow position region wrt left knee

FIGURE 5.23G

Kneeling.

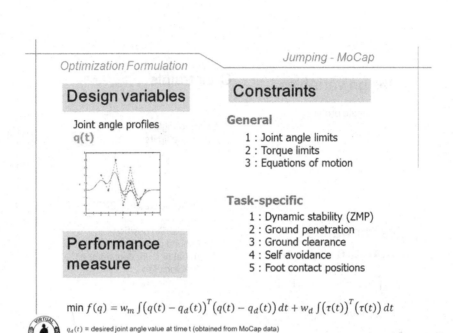

FIGURE 5.23H

Jumping.

where the dynamic equations of motion must be integrated, this method provides a viable and rigorous approach to solving for the motion.

The equations of motion are typically characterized by a set of differential algebraic equations and contain a high degree of freedom system. . .they are quite complex, and highly nonlinear. As a result, these DAEs are difficult to integrate if not impossible, depending on the complexity.

Predictive dynamics offers a unique method for characterizing human behavior as represented by cost functions (one or multiple objectives) formulated as a multi-objective optimization problem. A digital human climbing a wall will most likely have various cost functions driving that motion, including but not limited to minimizing energy, maximizing stability, maximizing vision, and several others. The predictive dynamics approach provides for a direct method to enable "human" or more "naturalistic" predictions of human motion than the typical direct integration of the equations of motion.

We list below the major important aspects of predictive dynamics:

a. No integration of the equations of motion.
b. Uses human cost functions to drive the motion.
c. Computationally efficient.

d. Can predict behavior and, more importantly, can simulate behavior by varying the selection of cost functions.

e. Can assess cause and effect...allows for tradeoff analysis.

References

Arora, J.S., Wang, Q., 2005. Review of formulations for structural and mechanical system optimization. Struct. Multidisciplinary Optim. 30, 251−272.

Cahalan, T.D., Johnson, M.E., Liu, S., Chao, E.Y.S., 1989. Quantitative measurements of hip strength in different age-groups. Clin. Orthop. Relat. Res. 246, 136−145.

Gill, P.E., Murray, W., Saunders, M.A., 2002. SNOPT: an SQP algorithm for large-scale constrained optimization. Siam J. Optim. 12 (4), 979−1006.

Gubina, F., Hemami, H., McGhee, R.B., 1974. On the dynamic stability of biped locomotion. Biomed. Eng. IEEE Trans. 2, 102−108.

Huang, Q., Yokoi, K., Kajita, S., Kaneko, K., Arai, H., Koyachi, N., et al., 2001. Planning walking patterns for a biped robot. IEEE Trans. Rob. Autom. 17 (3), 280−289.

Kaminski, T.W., Perrin, D.H., Gansneder, B.M., 1999. Eversion strength analysis of uninjured and functionally unstable ankles. J. Athl. Train. 34 (3), 239−245.

Kim, H.J., Horn, E., Arora, J.S., Abdel-Malek, K., 2005. An Optimization-Based Methodology to Predict Digital Human Gait Motion. SAE International, Warrendale, PA (SAE Paper No. 05DHM-54).

Kim, H.J., Wang, Q., Rahmatalla, S., Swan, C., Arora, J., Abdel-Malek, K., et al., 2008. Dynamic motion planning of 3D human locomotion using gradient-based optimization. J. Biomech. Eng. 130 (3), 031002-1−031002-14.

Kumar, S., 1996. Isolated planar trunk strengths measurement in normals 3. Results and database. Int. J. Ind. Ergon. 17 (2), 103−111.

Rasmussen, J., Damsgaard, M., Voigt, M., 2001. Muscle recruitment by the min/max criterion—a comparative numerical study. J. Biomech. 34 (3), 409−415.

Saunders, J.B.D.M., Inman, V.T., Eberhart, H.D., 1953. The major determinants in normal and pathological gait. J. Bone Joint Surg. Am. 35-A (3), 543−558.

Vukobratović, M., Borovac, B., 2004. Zero-moment point—35 years of its life. Int. J. HR 1 (1), 157−173.

Wang, Q., Xiang, Y., Kim, H., Arora, J., Abdel-Malek, K., 2005. Alternative Formulations for Optimization-Based Digital Human Motion Prediction. SAE International, Warrendale, PA (SAE Paper No. 05DHM-61).

Xiang, Y., 2008. Optimization-based dynamic human walking prediction, PhD. Dissertation, The University of Iowa, 141 pages.

Xiang, Y., Chung, H.J., Mathai, A., Rahmatalla, S., Kim, J., Marler, T., et al., 2007. Optimization-based dynamic human walking prediction. Paper presented at the SAE Digital Human Modeling Conference, Seattle, WA.

Xiang, Y., Arora, J.S., Abdel-Malek, K., 2009a. Optimization-based motion prediction of mechanical systems: sensitivity analysis. Struct. Multidisciplinary Optim. 37 (6), 595−608.

Xiang, Y., Arora, J.S., Rahmatalla, S., Abdel-Malek, K., 2009b. Optimization-based dynamic human walking prediction: one step formulation. Int. J. Numer. Methods Eng. 79 (6), 667−695.

Xiang, Y., Arora, J.S., Rahmatalla, S., Bhatt, R., Marler, T., Abdel-Malek, K., 2009c. Human lifting simulation using multi-objective optimization approach. Multibody Syst. Dyn. 23 (4), 431−451.

Xiang, Y., Arora, J.S., Abdel-Malek, K., 2010a. Optimization-based prediction of asymmetric human gait. J. Biomech. 44 (4), 683−693.

Xiang, Y., Arora, J.S., Abdel-Malek, K., 2010b. Physics-based modeling and simulation of human walking: a review of optimization-based and other approaches. Struct. Multidisciplinary Optim. 42 (1), 1−23.

Xiang, Y., Chung, H.J., Kim, J.H., Bhatt, R., Rahmatalla, S., Yang, J., et al., 2010c. Predictive dynamics: an optimization-based novel approach for human motion simulation. Struct. Multidisciplinary Optim. 41 (3), 465−479.

Strength and Fatigue: Experiments and Modeling

6

The majority of this chapter was contributed by Laura Frey-Law, PhD, MS, PT

The difference between a successful person and others is not a lack of strength, not a lack of knowledge, but rather a lack of will.
Vince Lombardi

6.1 Joint space

In order to determine whether a digital human model can accomplish a task, it is essential to define strength limits. The simple concept of measuring strength is indeed simple to understand, but difficult to attain. The intuitive approach is to consider muscle strength and forces as an assessment of strength for various regions of the anatomy. However, one quickly realizes that this is a difficult task as the concept of muscle recruitment and activation is difficult to ascertain.

This makes a seemingly simple modeling process somewhat more complex than one might initially expect. Clearly an individual who is "strong" can be readily differentiated from one who is "weak". However, there are several physiological aspects of muscle strength that may be relevant to digital humans, as well as the decay in strength (i.e., muscle fatigue) that occurs naturally with physical activity.

Reasonable representations of normative human strength and endurance are needed for digital human models (DHM) to behave in expected and meaningful ways. Without accurate models of strength (and fatigue) DHM cannot model typical or realistic human behavior reliably.

Models of muscle strength typically fall into two basic approaches: modeling individual muscle forces or modeling net torque produced about a joint, due to multiple synergistic muscles. Both modeling approaches have their advantages and disadvantages, particularly for whole-body digital human modeling. However, we will focus our attention on strength modeling at the joint level as that is the level at which we can assess strength in humans, and thus for validation. This is also called the joint space, as all calculations, predictions, and experimental measurements are done at the joint level.

Muscle model approaches require assumptions of load sharing and activation levels that are also pertinent to the issues discussed in this chapter. However, non-invasive direct muscle force measurements are not possible in humans; only net

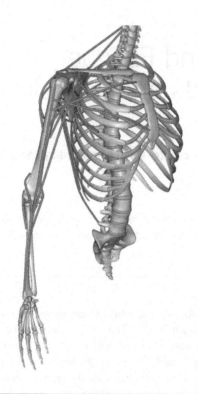

FIGURE 6.1

Muscle modeling using curves of action (Patrick, 2007).

torque measurements produced about a joint. Thus, all models of strength must ultimately be validated against assessments of net joint torque. A muscle-level model of a DHM is shown in Figure 6.1 and developed by the authors' team (Patrick, 2007) where muscle lines and curves of action are depicted. This model provides for muscle curves that wrap and slide.

If torque is calculated at the joint, which is at the heart of this book, then muscle forces can be calculated by resolving the torque into its various components (Figure 6.2). And indeed there are several methods for accomplishing this calculation (Patrick, 2007; Zatsiorsky, 2002).

6.2 Strength influences

Muscle strength definitions typically center on the capability to produce force, or the maximal force-production of a muscle. Several factors can influence muscle strength, including properties inherent to muscle contraction (i.e., at the cellular, fiber, or single muscle level) and the mechanical arrangement of muscles within the musculoskeletal system (i.e., considering individual and/or groups of muscles).

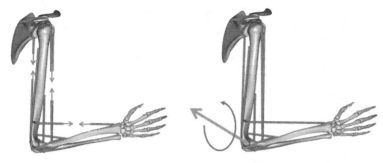

FIGURE 6.2

Resolving a torque at a joint to muscle forces.

At the single muscle level, strength is proportional to muscle cross-sectional area. This is commonly observed as the phenomenon of large, hypertrophied muscles being stronger than small, atrophied muscles. Total muscle length is not a critical factor in muscle strength per se, but may indirectly influence force-production through the mechanical advantage of the muscle or the ability of the muscle to contract rapidly (e.g., contraction velocity), which will be discussed below. Conversely, the relative length of the muscle influences force-production through the length-tension relationship, which will be discussed further below.

Muscle hypertrophy occurs as an adaptation to unaccustomed activity or strength training, and strength training is often performed precisely because of this relationship, namely to create larger muscles. The specific tension of muscle, that is the force produced per unit area, is relatively constant across muscles, men and women, and is often used to estimate peak force for single muscle models. Specific tensions have been reported to range from approximately $30\,\text{N/cm}^2$ (Chow et al., 1999; Narici et al., 1992; Reeves et al., 2004) to $55-60\,\text{N/cm}^2$ (O'Brien et al., 2010). The variations in estimates are likely due to differences in how physiological cross-sectional area and/or peak force are accounted for (e.g., fiber angle at rest or during contraction; antagonist muscle force considered, etc). However, specific tension does not vary between men and women or children versus adults (O'Brien et al., 2010). Thus, differences in strength between individuals are largely due to differences in the muscle cross-sectional area.

Another inherent muscle property influencing force-producing capability is the torque-velocity relationship (Figure 6.3). Hill (1938) first described this nonlinear phenomenon over 80 years ago. Essentially, muscles produce less peak force the faster they shorten (Hill, 1938). Practically, this is apparent when one tries to lift a light versus a heavy weight in one hand as fast as the muscles can. They are able to move quite quickly with a light hand weight (e.g., 2 lbs), but move much slower when trying to move a heavy weight (e.g., 20 lbs). Interestingly this occurs during shortening, or concentric, contractions, but the opposite occurs during lengthening, or eccentric, contractions (see Figure 6.3). Eccentric contractions occur when

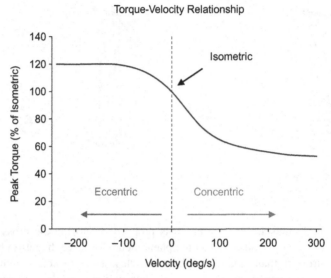

FIGURE 6.3

Schematic representation of the torque-velocity relationship for human muscle.

muscles are lengthened while resisting external loads. For example lowering an object, walking down stairs, or performing "negative" push-ups (i.e., the lowering phase of a push-up) all involve eccentric contractions. In animal studies eccentric contractions produced through electrical stimulation can produce up to 150% of peak isometric (static) contraction forces (Lieber, 2002); however, in humans voluntary eccentric contractions typically only reach levels between 100–120% of isometric peak torque (Chapman et al., 2005; 2008; Griffin, 1987; Horstmann et al., 1999; Klass et al., 2005; Kramer and Balsor, 1990).

A side note regarding eccentric contractions is that the common experience of developing muscle soreness 1–2 days after exercise (i.e., delayed onset muscle soreness, DOMS) has been shown to be a result of only the eccentric and not the isometric or concentric component of unaccustomed tasks (Cleak and Eston, 1992; Yu et al., 2002).

Two factors contribute to the influence of joint angle on muscle strength. First, the length of the contracting muscle fibers has a direct influence on muscle force produced (i.e., the length-tension relationship) and second, the varying moment arm through a joint range of motion has an indirect influence on muscle strength via mechanical advantage (i.e., torque = force × moment arm). Muscle contraction occurs as two myofilament proteins bind and slide past one another: the actin and myosin filaments. There is an optimal level of overlap between these filaments which results in peak force-production from a muscle fiber. However, if this overlap increases or decreases from the optimal length, the ability of the fiber to produce active muscle force decays, creating a curvilinear

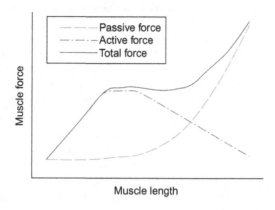

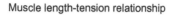

FIGURE 6.4

Schematic of the muscle length-tension relationship, demonstrating the active, passive and net tension produced as a function of muscle length.

force-length relationship (Gordon et al., 1966). In addition to this active force, muscle and tendon exhibit passive elastic properties, such that stretching a muscle eventually produces passive force, like a spring. If you combine these active and passive forces, the net force from a muscle typically increases with muscle length, but with a curvilinear plateau in the region of optimal muscle length (Figure 6.4).

The torque produced about a joint is the product of the muscle force and its moment arm with respect to the joint center of rotation. Thus, torque is a function of both muscle force-producing capability and the mechanical advantage afforded by the muscle-tendon complex. As joint angles change throughout the range of motion, this mechanical advantage changes. These changes can be specific to each joint, thus no one relationship can be modeled for all joints (Frey-Law et al., 2012b). Thus, the torque-angle relationship does not always appear to be consistent with the force-length relationship.

Population-specific factors that influence strength include: male versus female; young-versus old-adult; and active/trained versus sedentary cohorts, to name a few. Typically men exhibit approximately 50% greater peak torque than females but this too can vary across joint regions (Frey-Law et al., 2012b). As mentioned above, these sex differences result primarily from cross-sectional area as opposed to the inherent muscle properties for men and women (e.g., specific tension of muscle). Certainly this does not discount systemic physiological differences between men and women, including sex differences in the hormonal milieu, from playing a strong role in muscle strength. It does, however, suggest the contractile proteins are not in and of themselves the source of these differences per se, but rather the number of these muscle fibers in parallel. Accordingly, digital humans (DHs) need to be specific to male versus female avatars not only in physical anthropometry (e.g., height and weight) but also in strength properties.

Strength is considered to be relatively stable from the third (i.e., 20 s) to the fifth decades of life (i.e., 40 s), but typically begins to decay sometime after the sixth decade of life (i.e., in 50 s or 60 s) (Stoll et al., 2000). This suggests DH models of strength can be used to represent "young" adults across several decades of age. Whereas others have suggested using linear decays in strength models from the 20 s to the 70 s, data suggest constant strength models through the 40 s or 50 s are reasonably well substantiated (Bohannon, 1997; Horstmann et al., 1999; Stoll et al., 2000). Future advances in DH modeling may move toward more exact models of strength, by decade of life, but this level of fidelity does not currently have sufficient data to suggest its usefulness. However, DHs representing the older adult population (i.e., over 65 years) clearly would benefit from strength models specific to this cohort. This loss of strength with increasing age is likely due to a complex array of factors such as relative inactivity, sub-threshold disease or pathology, or age-related changes in cell health and adaptation. While age-related declines in health can vary between men and women or between muscle groups, one simple modeling approach is to reduce young-adult strength by a constant percentage (e.g., 15%), which could also be increased with advancing age.

Changes in strength with training can be quite large and specific to the muscle groups involved. While this is not modeled directly by most DH strength models today, it can be represented indirectly using population statistics. Most normative data collected on strength will include a heterogeneous population, resulting in observed standard deviations in the range of 20–30% of mean strength values (Bohannon, 1997; Frey-Law et al., 2012b; Griffin, 1987; Horstmann, Maschmann et al., 1999). Thus, these data sets can be used to model stronger or more highly trained individuals using the mean plus some multiple of the observed standard deviation (i.e., higher strength population percentiles). Conversely, untrained or weak individuals can be modeled using the observed mean minus some multiple of the standard deviation (i.e., lower strength population percentiles). Direct models attempting to link training and strength are not likely to be practical anytime soon, however, as training effects are often non-linear, occurring in proportion to the baseline level of training, frequency and intensity of training, etc. For example, individuals that are sedentary will see larger adaptations with training than someone who is already near their peak performance level.

6.3 Strength assessment

The actual assessment of strength can be accomplished in several ways, each measuring different aspects of muscle strength capability. Isometric (i.e., constant length, static) assessments are performed at specific joint angles and typically used to evaluate torque-angle relationships. These measurements are often assessed using isokinetic dynamometers (set to work isometrically) or load cells (force assessments at a constant moment arm length). Isotonic (i.e., constant load)

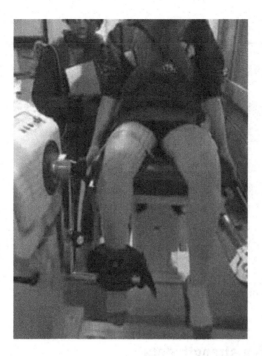

FIGURE 6.5

A subject conducting a knee strength measurement using an isokinetic dynamometer.

assessments either use a constant mass (e.g., free weights) and are assessed as the one repetition maximum (1 RM) that an individual can lift through a full ROM, or use an isokinetic dynamometer set to maintain a constant external torque, allowing the contraction velocity to vary throughout the ROM. Isokinetic (i.e., constant velocity) assessments also typically use an isokinetic dynamometer, but with a different setting; using a constant velocity, the torque produced throughout the ROM is measured. In research settings, the most common means to assess strength involve isokinetic dynamometers, due to their flexibility in strength assessments, followed by load cell configurations; whereas isotonic assessments are largely used in fitness and health centers due to their relatively low cost (the assessment can be carried out using only a set of free weights).

Figures 6.5 and 6.6 show subjects conducting strength measurements using an isokinetic strength dynamometer. Isokinetic exercise and testing are tools used to develop and assess strength of the various joints such as the elbow, knee, wrist, shoulder, hip, and torso. These machines or devices typically provide a computer-controlled torque motor or brake to limit the maximum movement velocity and thus obtain isokinetic motion during maximum effort trials. The force the subject applies to the device is measured and recorded along with joint angle and movement velocity, which then can be used to assess peak strength for a given isokinetic velocity. Concentric, eccentric, and isometric modes of exercise can be used.

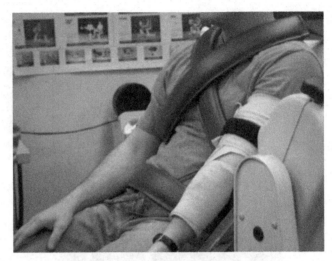

FIGURE 6.6

Strength assessment for the elbow using an isokinetic dynamometer.

6.4 Normative strength data

There are numerous studies reporting on joint-level strength, particularly focusing on torque-angle (isometric) and/or torque-velocity (isokinetic) data. Advances in DHM coupled with the use of predictive dynamics or optimization techniques can better utilize information gleaned from both of these relationships more so than previous quasi-static DH models, which relied almost solely on torque-angle data.

To best utilize both of these "two-dimensional" curvilinear relationships (i.e., 2D curves of torque-angle or torque-velocity), recently studies have begun to evaluate "three-dimensional" strength relationships: torque as a function of both velocity and angle, creating 3D surfaces (Anderson et al., 2007; Frey-Law et al., 2012b; Khalaf and Parnianpour, 2001; Khalaf et al., 1997, 2000, 2001). For example, the strength surfaces shown in Figure 6.7 represent the mean concentric strength values for knee flexion and extension torque as a function of joint angle and movement velocity (i.e., 3D strength surfaces), for both males and females.

Note that the shape of the flexion and extension surfaces differ, suggesting the mechanical advantages due to muscle lengths and muscle moment arms play a significant role in determining the 3D peak torque surface. However, the surfaces are qualitatively similar between men and women; they differ primarily in absolute strength levels. Similarly, 3D strength surfaces for shoulder flexion and extension are shown in Figure 6.8.

These 3D surfaces have been modeled in various ways, including modifications to the torque-velocity relationship accounting for changes in joint angle and eccentric contractions (Anderson et al., 2007) and second-order polynomials

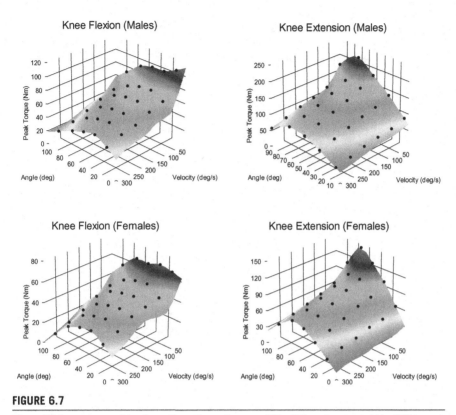

FIGURE 6.7

Mean 3D knee concentric strength surfaces for men ($N = 24$) and women ($N = 21$).

(Khalaf and Parnianpour, 2001; Khalaf et al., 1997, 2000, 2001). While both of these approaches provide reasonable estimates of the peak strength surfaces, they do not readily allow for significant interactions between joint angle and contraction velocity which are observed in experimental data (Frey-Law et al., 2012b), and can result in non-physiologic torque predictions (i.e., values "crossing zero") when extrapolated to joint angles beyond those assessed for developing the strength models.

We have found greater flexibility using logistic equations to best fit the nonlinear experimental data. This approach maximizes the heuristic, nonlinear representation of strength surfaces using only 7 to 8 parameters, thus is also reasonably parsimonious. We chose to model the eccentric relationship as it is challenging to train individuals to exert maximum eccentric strength and can lead to DOMS or more severe muscle injury. Based on numerous previous studies, we modeled peak eccentric strength as 120% of peak isometric strength independent of eccentric contraction velocity. Using this assumption, examples of the resulting 3D surfaces that model both concentric and eccentric

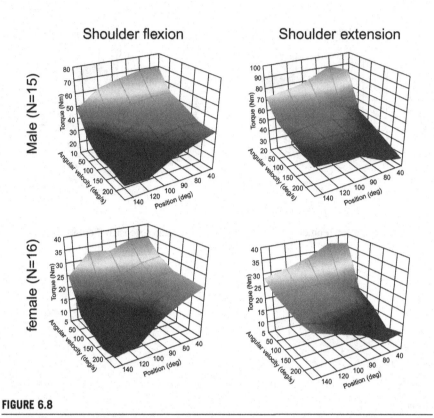

FIGURE 6.8

Strength surfaces for shoulder flexion and extension torque for males and females.

strength using logistic equations are shown in Figure 6.9. The format of this approach is provided in Equation (6.1), where X = joint angle, Y = contraction velocity, Z = peak torque, and parameters a–h define a unique surface for each joint and torque direction:

$$Z = a + \text{LOGISTIC X (b, c, d)} + \text{LOGISTIC Y (e, f, g)}$$
$$+ \text{LOGISTIC X (h, c, d)} * \text{LOGISTIC Y (1, f, g)}$$

(6.1)

However, specific parameter values will depend on the units used (e.g., degrees vs radians) and the sign conventions needed for any particular digital human. That is, we display all strength values as positive here, but for any specific DHM application the joint angles, contraction velocities, and torque signs have to be adjusted to match the underlying framework of the model (e.g., knee flexion torque may be negative and knee extension torque positive, with the joint angles represented as negative values as the knee moves into flexion). Thus, no one set of parameter values is likely to be universal across all DH models.

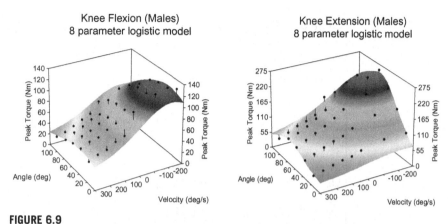

FIGURE 6.9

Measured (concentric, from Figure 6.3) and estimated (eccentric, 120% of isometric peak torque) strength plotted as points, with the logistic, 8-parameter surface models shown concurrently.

6.5 Representing strength percentiles

Clearly, strength varies across individuals, thus normative strength can be best represented by population means and/or strength percentiles. These strength surfaces are frequently fit from average or median data, however could also be fit to any strength population percentile (%ile). For DHM, normative strength can be easily modeled using z-scores, where any percentile level is equal to the mean plus the appropriate z-score (see Table 6.1 for standard z-scores by percentile) multiplied by the standard deviation (SD, Equation 6.2). Thus, the 50th %ile strength level for either men or women is simply the mean (since $z = 0$ for 50th %ile). This is valid for normally distributed data, which we found to be true for knee and elbow strength (Frey-Law et al, 2012b) when separated by sex.

$$\%ile\ strength = mean\ strength + (z\text{-score }\%ile * SD) \tag{6.2}$$

The standard deviation can also be modeled as a 3D surface as a function of contraction velocity and joint angle, but a simpler approach is to represent it by the mean coefficient of variation (CV, Equation 6.3) across all joint angles and contraction velocities. The coefficient of variation is simply the standard deviation standardized by the mean. This allows for limited modulation of standard deviation values with little additional computational effort. We have found CV varies somewhat between joints and contraction types but centers roughly on a value of 0.3 (Frey-Law et al., 2012b).

$$CV = SD/mean \tag{6.3}$$

Thus, combining Equations (6.2) and (6.3), we can calculate strength percentiles as indicated in Equation (6.4) (i.e., %ile strength is a function of joint

Table 6.1 Standard z-scores for Percentile Calculations (Normal distribution)

Percentile (%ile)	Z-score
1	−2.330
5	1.645
10	−1.280
15	−1.040
20	−0.840
25	−0.675
30	−0.525
35	−0.385
40	−0.250
45	−0.125
50	0.000
55	0.125
60	0.250
65	0.385
70	0.525
75	0.675
80	0.840
85	1.040
90	1.280
95	1.645
99	2.330

angle (q) and contraction velocity (dq/dt)). The 3D surface models of peak torque are used to provide the mean peak torque values for each joint angle (q) and contraction velocity (dq/dt) of interest.

$$\%\text{ile strength}\,(q, dq/dt) = \text{mean}\,(q, dq/dt) + (z\text{-score}^{*}\,CV^{*}\,\text{mean strength}\,(q, dq/dt))$$
$$= (1 + z\text{-score}^{*}\,CV)^{*}\,\text{mean}\,(q, dq/dt)$$

(6.4)

where CV is a joint-specific constant.

6.6 Mapping strength to digital humans: strength surfaces

The mapping of normative joint strength to digital humans is not necessarily a trivial matter and depends on the DH framework used and the anatomical definition for joint range of motion. For relatively simple hinge joints, such as the knee or elbow, this

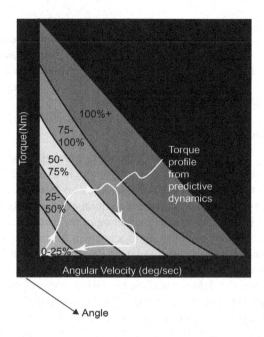

FIGURE 6.10

Torque-velocity results from predictive dynamics are compared with extrapolated torque-velocity curves (surfaces).

process may be relatively straightforward involving simply confirming whether a "straight" knee is defined as 0 or 180 degrees and matching the sign conventions between the DH model and the normative data. However, for 3-DOF joints (or more), such as the hip, trunk, or shoulder, this process is complicated by computational issues, such as gimbal lock, and measurement issues such as separating net joint torque into the underlying DH joints. For example, shoulder torque is typically assessed as net flexion/extension, abduction/adduction, internal/external rotation, and possibly horizontal abduction/adduction. However, these torques occur due to muscular torques about both the glenuhumeral and scapulothoracic joints.

Depending on the DH model, it may be challenging to discern which joints should be compared with which normative data values, or if some load sharing paradigm must be assumed between the joints involved. Similarly, trunk strength is frequently assessed as a single entity in physiological measurements (e.g., net trunk extension torque), yet clearly there are multiple spinal joints involved. However, the muscles involved span several joints and cannot be isolated to only a single joint (L4-L5 verses L5-S1, for example). Thus, assumptions must be made to enable strength mapping between experimental, normative data and DHM, and errors associated with gimbal lock must be noted.

Figure 6.10 demonstrates the concept of comparing torque values calculated from predictive dynamics with those measured experimentally using a 2D torque-velocity curve for simplicity. The white curve depicts torque values over time for a single joint relative to expected torque-velocity percentile strength curves. Note that for this example, the task requires from <25% of maximum to >75% of maximum peak torque, which varies across time.

Once strength models have been created, there are essentially two methods for applying torque models in DHM:

a. Pre-processing approach: Embedding the torque/strength models as limits or constraints in the optimization algorithm which yields a motion that is constrained to the available peak strength, i.e., a motion that can be accomplished by the digital human if a solution exists. The benefit of this method is that it simulates human motion more realistically as strength limits (e.g., 50th %ile) cannot be violated to accomplish the motion, if the task can be accomplished. However, the disadvantages include greater computation cost (i.e., a nonlinear constraint that varies as a function of the parameters being optimized) and it does not allow the DHM to model where a task could exceed the strength surface for any given strength percentile, which may be useful for predicting risks for musculoskeletal injury.

b. Post-processing approach: Comparing the results of the predicted DHM torques resulting from the optimization algorithm using only simplified strength limits (e.g., a constant average value) to the 3D strength surfaces without using the surface models to limit or constrain the DHM. This method is computationally simpler and provides insight into what joint has exerted a torque that may exceed expected normative percentile levels. However, it can also yield motions that would otherwise not be possible, particularly at faster velocities or end-ranges of motion where the simple strength limit is most likely to be in error relative to the 3D strength models.

6.7 Fatigue

Localized muscle fatigue can be briefly defined as the loss of force-producing capability following muscle activity (Bigland-Ritchie and Woods, 1984). Thus, fatigue is temporary, recovers following rest, and is distinct from weakness, pathology or traumatic injury. Practically, the development of fatigue is important for dynamic DHM because the temporal decay in muscle strength is a very common and real phenomenon. Clearly, as we perform tasks either for greater lengths of time or of higher intensity, we typically develop greater fatigue.

A curvilinear relationship between fatigue and intensity has been well documented for over 50 years, often referred to as Rhomert's curve or the intensity-endurance time (ET) curve (Rohmert, 1960). Numerous authors have proposed various versions of intensity-ET curves, including several joint-specific models (El ahrache et al., 2006;

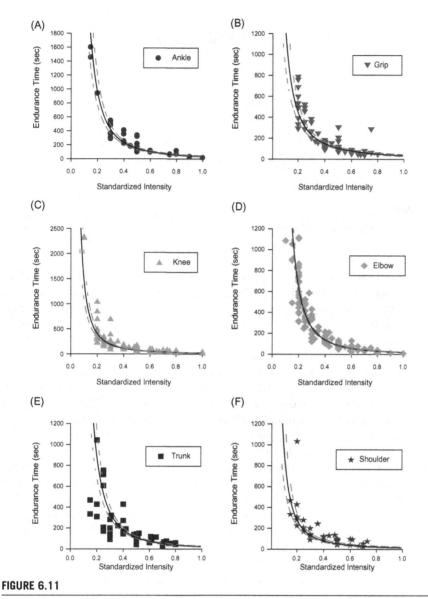

FIGURE 6.11

Joint-specific statistical fatigue models (mean ±95% prediction intervals) and their corresponding experimental data points from the literature.

Reprinted with permission from Frey-Law and Avin, 2010

Monod and Scherrer, 1965; Rose et al., 2000) based on limited fatigue data and/or a compilation of other models. We recently performed a systematic review of the literature to gather all relevant static fatigue data (194 publications were included, with 369

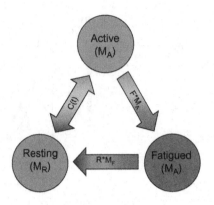

FIGURE 6.12

Schematic representation of the three-compartment fatigue model.

Reprinted with permission from Frey-Law et al., 2012a

data points) and generated new intensity-ET curves for a "general" model as well as six joint-specific models (Figure 6.11): hand/grip, elbow, shoulder, trunk, knee, and ankle (Frey-Law and Avin, 2010). These statistical models provide a simplified means to estimate the development of fatigue (e.g., by estimating maximum endurance time) for digital human applications.

These intensity-ET relationships are particularly relevant for static, sustained tasks with no rest intervals; however, most activities involve some component of dynamic movement and/or relative rests. Thus analytical, heuristic models which can represent the actual changes in peak force-production are needed to represent the decay and recovery of muscle fatigue at the joint-level for DHM applications.

Several models have been proposed to model fatigue, but relatively few are specifically for DH applications. Ding and colleagues developed a muscle fatigue model that was aimed at clinical applications for electrical stimulation of paralyzed muscle (Ding et al., 2000, 2002a,b, 2003). This modeling approach is an adaptation to a muscle model by adding a decay component and requires estimates of stimulation frequency input to the motor neurons. We developed a different approach to model fatigue which is capable of addressing the effects of work intensity, joint position, contraction velocity, and rest intervals (Xia and Frey-Law, 2008a,b).

Our model (Figure 6.12) was adapted from a previous model first proposed for modeling only maximal contractions (Liu et al., 2002), employing a three-compartmental approach. Muscle tissues are considered to be in one of three states (i.e., three compartments) at any time: resting (M_R), activated (M_A), or fatigued (M_F). We added a feedback control loop to the model, allowing for submaximal contractions to be modeled. Muscle activation-deactivation ($M_R \leftrightarrow M_A$) is simulated using a bounded proportional controller to regulate (i.e., match) required work intensity (as predicted by a DHM). The fatigue ($M_A \rightarrow M_F$) and recovery ($M_F \rightarrow M_R$) processes are regulated by two transfer rate coefficients,

F and R, representing the fatigue and recovery properties, respectively. The system can maintain a steady torque output by increasing muscle activation until failure, i.e. maximum holding time (MHT), occurs. Additionally, the model can calculate the level of fatigue at any given time during a task as the proportion of the muscle in the "fatigued" state (i.e., in M_F).

The model consists of several differential equations defining the flow rates between compartments, where the rate is proportional to the concentration (volume) of each compartment (see Equations 6.5−6.7) and an equation for the controller depending on the situation (Equations 6.8−6.10). The input required is the relative task intensity, which may simply be known (i.e., for a simple task) or can be modeled using DHM to estimate required net joint torques standardized by the 3D strength surfaces (as a function of joint angle and velocity). Thus, in this way, even a complex dynamic task can be represented as a time-vector of task intensity values (i.e., ranging in value from 1−100% intensity) as previously depicted schematically in Figure 6.10. We can represent joint-specific fatigue behavior through variations in the model parameters, F and R:

$$dM_R/dt = -C(t) + R * M_F \qquad (6.5)$$

$$dM_A/dt = C(t) - F * M_A \qquad (6.6)$$

$$dM_F/dt = F * M_A - R * M_F \qquad (6.7)$$

$$C(t) = L * (TL - M_A) \qquad (6.8)$$

i.e., activation of muscle, if there is sufficient "muscle" available in the active and/or resting compartment to meet the target level requirements

$$C(t) = L * M_R \qquad (6.9)$$

i.e., activation of muscle, if there is insufficient "muscle" available in the active and/or resting compartment to meet the target level requirements

$$C(t) = L * (TL - M_A) \qquad (6.10)$$

i.e., deactivation of muscle, if the active compartment is exceeding the target level

where

$C(t)$ = the controller denoting the muscle activation-deactivation drive;
F = fatigue parameter defining the rate of change between the active and fatigued compartments;
R = recovery parameter defining the rate of change between the fatigued and resting compartments;
L = an arbitrary constant tracking factor to ensure good system behavior (Xia and Frey-Law, 2008a). We found a value of 10 is reasonable based on a sensitivity analysis (Xia and Frey-Law, 2008c); and

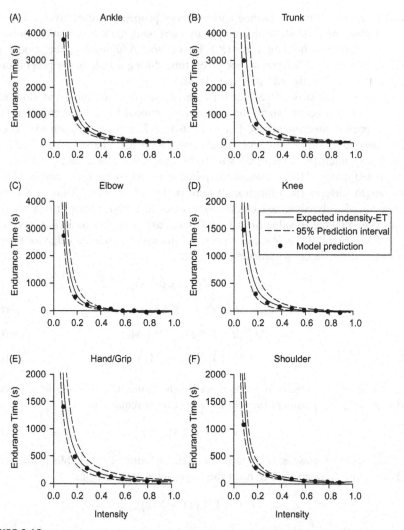

FIGURE 6.13

Model predictions compared with statistical fatigue model (Frey-Law and Avin, 2010). Note how well the three-compartment model can reproduce the static intensity-ET relationships for each joint, simply using two model parameters.

Reprinted with permission from Frey-Law et al., 2012a

TL = target level for the task (% maximum) that the controller is attempting to match.

We have validated the model against statistical fatigue models obtained from our meta-analysis systematic review (Frey-Law et al., 2012a). This approach reproduces the intensity-ET curves exceedingly well (Figure 6.13) and is mathematically capable of predicting intermittent tasks as well as dynamic tasks for use with predictive dynamics.

6.8 Strength and fatigue interaction

One method to combine strength and fatigue models for DHM is to use the fatigue model to decay the corresponding strength surface (e.g., joint-specific). For example, as the elbow flexors are used (task modeled in DHM to provide task intensity input to fatigue model), the fatigue model can then predict the decay in peak strength capability over time. This can be conceptualized as a fatigue coefficient with values from 0 (completely fatigued) to 1 (completely rested). The product of this coefficient with the peak 3D strength surfaces results in the time-varying strength properties of a joint as a function of time (and the underlying task involved).

Continued efforts are needed to validate and determine the practical applications of these modeling approaches, but significant advances in modeling strength and fatigue have occurred in the past decade. Future advances are likely to outpace previous accomplishments, making this area an exciting component in the development of DHM tools.

6.9 Concluding remarks

In summary, representing strength and fatigue for digital human models can be accomplished with surprising accuracy at the joint level, which we call the joint space. By applying both well-documented musculoskeletal relationships with new data and modeling approaches, numerous muscle force nonlinearities can be efficiently modeled computationally.

Modeling muscle forces as the limiting factor for a particular motion is both difficult and impractical, whereas joint-space strength limits provide a computationally efficient approach. Predicted joint torques obtained from predictive dynamics can readily be compared against normative torques (strength percentile surfaces, postprocessing) or the 3D strength surfaces can be used as constraints (pre-processing), which better limit the capability of a human model to perform a task.

Undoubtedly, future advances will continue to improve the accuracy of these modeling approaches, making digital humans increasingly realistic and useful tools for a wide variety of applications.

References

Anderson, D.E., Madigan, M.L., Nussbaum, M.A., 2007. Maximum voluntary joint torque as a function of joint angle and angular velocity: model development and application to the lower limb. J. Biomech. 40 (14), 3105–3113.

Bigland-Ritchie, B., Woods, J.J., 1984. Changes in muscle contractile properties and neural control during human muscular fatigue. Muscle Nerve 7 (9), 691–699.

Bohannon, R.W., 1997. Reference values for extremity muscle strength obtained by hand-held dynamometry from adults aged 20 to 79 years. Arch. Phys. Med. Rehabil. 78 (1), 26–32.

Chapman, D., Newton, M.J., Nosaka, K., 2005. Eccentric torque-velocity relationship of the elbow flexors. Isokinet. Exerc. Sci. 13 (2), 139−145.

Chapman, D.W., Newton, M.J., Zainuddin, Z., Sacco, P., Nosaka, K., 2008. Work and peak torque during eccentric exercise do not predict changes in markers of muscle damage. Br. J. Sports Med. 42 (7), 585−591.

Chow, J.W., Darling, W.G., Hay, J.G., Andrews, J.G., 1999. Determining the force-length-velocity relations of the quadriceps muscles: III. A pilot study. J. Appl. Biomech. 15 (2), 200−209.

Cleak, M.J., Eston, R.G., 1992. Delayed onset muscle soreness: Mechanisms and management. J. Sports Sci. 10 (4), 325−341.

Ding, J., Wexler, A.S., Binder-Macleod, S.A., 2000. A predictive model of fatigue in human skeletal muscles. J. Appl. Physiol. 89 (4), 1322−1332.

Ding, J., Wexler, A.S., Binder-Macleod, S.A., 2002a. A predictive fatigue model−I: predicting the effect of stimulation frequency and pattern on fatigue [erratum appears in IEEE Trans Neural Syst Rehabil Eng. 2003 Mar; 11(1):86]. IEEE Trans. Neural. Syst. Rehabil. Eng. 10 (1), 48−58.

Ding, J., Wexler, A.S., Binder-Macleod, S.A., 2002b. A predictive fatigue model−II: predicting the effect of resting times on fatigue. IEEE Trans. Neural. Syst. Rehabil. Eng. 10 (1), 59−67.

Ding, J., Wexler, A.S., Binder-Macleod, S.A., 2003. Mathematical models for fatigue minimization during functional electrical stimulation. J. Electromyogr. Kinesiol. 13 (6), 575−588.

El ahrache, K., Imbeau, D., Farbos, B., 2006. Percentile values for determining maximum endurance times for static muscular work. Int. J. Ind. Ergon. 36, 99−108.

Frey-Law, L.A., Avin, K.G., 2010. Endurance time is joint-specific: a modelling and meta-analysis investigation. Ergonomics 53 (1), 109−129.

Frey-Law, L.A., Looft, J., Heitsman, J., 2012a. A three-compartment muscle fatigue model accurately predicts joint-specific maximum endurance times for sustained isometric tasks. J. Biomech. 45 (10), 1803−1808.

Frey-Law, L.A., Laake, A., Avin, K.G., Heitsman, J., Marler, T., Abdel-Malek, K., 2012b. Knee and Elbow 3D strength surfaces: peak torque-angle-velocity relationships. J. Appl. Biomech. 28 (6), 726−737.

Gordon, A.M., Huxley, A.F., Julian, F.J., 1966. Variation in isometric tension with sarcomere length in vertebrate muscle fibres. J. Physiology−London 184 (1), 170−192.

Griffin, J.W., 1987. Differences in elbow flexion torque measured concentrically, eccentrically, and isometrically. Phys. Ther. 67 (8), 1205−1208.

Hill, A.V., 1938. The heat of shortening and the dynamic constants of muscle. Proc. R. Soc. Lond. Ser. B-Biol. Sci. 126 (843), 136−195.

Horstmann, T., Maschmann, J., Mayer, F., Heitkamp, H.C., Handel, M., Dickhuth, H.H., 1999. The influence of age on isokinetic torque of the upper and lower leg musculature in sedentary men. Int. J. Sports Med. 20 (6), 362−367.

Khalaf, K.A., Parnianpour, M., 2001. A normative database of isokinetic upper-extremity joint strengths: towards the evaluation of dynamic human performance. Biomed. Eng. App. Basis. Comm. 13, 79−92.

Khalaf, K.A., Parnianpour, M., Sparto, P.J., Simon, S.R., 1997. Modeling of functional trunk muscle performance: Interfacing ergonomics and spine rehabilitation in response to the ADA. J. Rehab. Res. Develop. 34 (4), 459−469.

Khalaf, K.A., Parnianpour, M., Karakostas, T., 2000. Surface responses of maximum isokinetic ankle torque generation capability. J. Appl. Biomech. 16 (1), 52–59.

Khalaf, K.A., Parnianpour, M., Karakostas, T., 2001. Three-dimensional surface representation of knee and hip joint torque capability. Biomed. Eng. App. Basis. Comm. 13, 53–56.

Klass, M., Baudry, S., Duchateau, J., 2005. Aging does not affect voluntary activation of the ankle dorsiflexors during isometric, concentric, and eccentric contractions. J. Appl. Physiol. 99 (1), 31–38.

Kramer, J.F., Balsor, B.E., 1990. Lower-extremity preference and knee extensor torques in intercollegiate soccer players. Can. J. Sport. Sci. (Revue Canadienne Des Sciences Du Sport) 15 (3), 180–184.

Lieber, R.L., 2002. Skeletal Muscle Structure, Function, and Plasticity. Lippincott Williams and Wilkins, Baltimore.

Liu, J.Z., Brown, R.W., Yue, G.H., 2002. A dynamical model of muscle activation, fatigue, and recovery. Biophys. J. 82 (5), 2344–2359.

Monod, H., Scherrer, J., 1965. The work capacity of a synergistic muscle group. Ergonomics 8, 329–338.

Narici, M.V., Landoni, L, Minetti, A.E., 1992. Assessment of human knee extensor muscles stress from *in vivo* physiological cross-sectional area and strength measurements. Eur. J. Appl. Physiol. Occup. Physiol. 65 (5), 438–444.

O'Brien, T.D., Reeves, N.D., Baltzopoulos, V., Jones, D.A., Maganaris, C.N., 2010. *In vivo* measurements of muscle-specific tension in adults and children. Exp. Physiol. 95 (1), 202–210.

Patrick, A., 2007. Development of a 3D Model of the Human Arm for Real-Time Interaction and Muscle Activation Prediction (MS Thesis), The University of Iowa, Iowa City, IA.

Reeves, N.D., Narici, M.V., Maganaris, C.N., 2004. Effect of resistance training on skeletal muscle-specific force in elderly humans. J. Appl. Physiol. 96 (3), 885–892.

Rohmert, W., 1960. Ermittlung von erholungspausen für statische arbeit des menschen (Determination of relaxation breaks for static work of man). Int. Z. Angew. Physiol. Einschl. Arbeitsphysiol. 18, 123–164.

Rose, L., Ericson, M., Ortengren, R., 2000. Endurance time, pain and resumption in passive loading of the elbow joint. Ergonomics 43 (3), 405–420.

Stoll, T., Huber, E., Seifert, B., Michel, B.A., Stucki, G., 2000. Maximal isometric muscle strength: normative values and gender-specific relation to age. Clin. Rheumatol. 19 (2), 105–113.

Xia, T., Frey-Law, L.A., 2008a. Multiscale Approach to Muscle Fatigue Modeling. Paper presented at the Pacific Symposium on Biocomputation, Big Island, HI.

Xia, T., Frey-Law, L.A., 2008b. Modeling Muscle Fatigue for Multiple Joints. Paper presented at the North American Congress on Biomechanics (NACOB), Ann Arbor, MI.

Xia, T., Frey-Law, L.A., 2008c. A theoretical approach for modeling peripheral muscle fatigue and recovery. J. Biomech. 41 (14), 3046–3052.

Yu, J.G., Malm, C., Thornell, L.E., 2002. Eccentric contractions leading to DOMS do not cause loss of desmin nor fibre necrosis in human muscle. Histochem. Cell Biol. 118 (1), 29–34.

Zatsiorsky, V.M., 1998. Kinematics of Human Motion. Human Kinetics Publishers, Champaign, Illinois.

Predicting the Biomechanics of Walking

7

With contribution by Yujiang Xiang, PhD, Virtual Soldier Research Program

If everything seems under control, you're just not going fast enough.
Mario Andretti

7.1 Introduction

The objective of this chapter is to develop a broadly applicable formulation for the prediction of human walking. It is to answer the question: "How would a person walk given a set of conditions, such as a specific body size, a specific load, and joint ranges of motion?"

Past research has addressed walking analysis, also called gait analysis, where a person's walk is measured and sometimes mathematically modeled for the reasons of analysis, to better understand deviation from the normal or to evaluate performance. This is of importance for the understanding of disabilities, injuries, and post-surgical gait, for example. The intent of this chapter is to create a predictive model for the biomechanics of walking.

Walking is also called ambulation and is one of the main gaits of locomotion among legged animals. Walking is generally defined by an inverted pendulum gait in which the body vaults over the stiff limb or limbs with each step. This applies regardless of the number of limbs—even for arthropods with six, eight or more limbs.

There are many reports that have established methods to simulate human motion, in particular walking, using databases generated from experiments on human subjects (Choi et al., 2003; Pettre and Laumond, 2006). These methods draw upon libraries of pre-stored motions to present the best-fit approach for a particular scenario. These approaches are limited by the accuracy and amount of available experimental data.

Other methods have attempted to model and simulate human walking. One such approach is to solve the walking problem based on the idea that biped walking can be treated as an inverted pendulum. Advantages of this method are its simplicity and faster solvable dynamics equations (Park and Kim, 1998). However, the method also suffers from an inadequate dynamics model that cannot generate natural and realistic human motion; particularly problematic is its inability to represent a large DOF system with high fidelity.

One well-established approach is the ZMP-based trajectory generation method, which has received a great deal of attention. In this method, the walking motion can be generated in real time for smaller-sized models to follow the desired ZMP trajectory using an optimal control approach (Kajita et al., 2003; Yamaguchi et al., 1999). The ZMP concept can also be incorporated in an optimization formulation to synthesize walking pattern by maximizing the stability, subject to physical constraints (Huang et al., 2001; Mu and Wu, 2003). The key point of this approach is that the dynamics equations are used only to formulate the stability condition rather than to generate the entire motion trajectory directly, thus many dynamics details are not considered.

Optimization-based trajectory generation presents a viable method for addressing the prediction of walking. This method has exhibited the possibility for predicting realistic and natural human motion. Furthermore, the method can easily handle large DOF models, and can optimize many human-related performance measures simultaneously and satisfy all the constraints (Anderson and Pandy, 2001; Chevallereau and Aoustin, 2001; Fregly et al., 2007; Lo et al., 2002; Ren et al., 2007; Saidouni and Bessonnet, 2003).

Chevallereau and Aoustin (2001) planned robotic walking and running motions using optimization to determine the coefficients of a polynomial approximation for profiles of the pelvis translations and joint angle rotations. Saidouni and Bessonnet (2003) used optimization to solve for cyclic, symmetric gait motion of a 9-DOF model that moves in the sagittal plane; the control points for the B-spline curves along with the time durations for the gait stages were optimized to minimize the actuating torque energy. Anderson and Pandy (2001) developed a musculoskeletal model with 23 DOFs and 54 muscles for normal symmetric walking on level ground. Muscle forces were treated as design variables and metabolic energy expenditure per unit distance was minimized. Lo et al. (2002) determined human motion that minimized the summation of the squares of all actuating torques. The design variables were the control points for the cubic B-spline approximation of joint angle profiles. Sensitivity of joint torque with respect to control points were analytically obtained by using the recursive Newton–Euler formulation.

Our team (Kim et al., 2005) has developed an optimization-based approach for predicting 3D human gait motions on level and inclined planes. By minimizing the deviation of the trunk from the upright posture, joint profiles were calculated, subjected to some physical constraints. Time durations for various gait phases were also optimized. The joint torques and ground reaction forces were not calculated; therefore constraints on the joint strength could not be imposed. In this chapter, recursive Lagrangian formulation is used for dynamics; joint torques and GRFs are calculated using equations of motion; a more realistic skeletal model is used; the optimization formulation is physics-based where an energy-related objective function is minimized; and constraints on the joint torques are imposed. As a result, a more realistic human walking motion is obtained.

It is noted that the material for this chapter is derived from several papers published by the authors and the contributing author to this chapter (Xiang, 2008; Xiang et al., 2009a,b, 2010a,b,c).

7.2 Joints as degrees of freedom (DOF)

To simulate walking, the lower part of the body is most important. The foot, ankle, knee, and hip joints are most critical for simulation and prediction of walking motion. However, including the spine and arms is also important because the inertia and dynamics associated with the upper torso and arms are significant to the balance and dynamics of the motion.

We shall use a 55-DOF human skeleton as presented in Chapter 2. There are 6 DOF for global translation and rotation, and 49 DOF representing the kinematics of the body. Each DOF corresponds to relative rotation of two body segments connected by a revolute joint.

We shall also use the Denavit−Hartenberg (DH) method as presented in Chapter 2. The objective is to calculate the motion of each of those DOF for each joint. Indeed, we also seek to calculate the torques required at each joint, denoted by torque profiles. While both joint and torque profiles are smooth curves that must be calculated for each joint, these curves will be represented with B-spline interpolations. As such, it is only the control points of each curve that must be calculated.

For dynamics and in order to represent the equations of motion, we shall use the recursive Lagrangian formulation which is known for its computational efficiency.

7.3 Muscle versus joint space

In the past, the emphasis in predicting motion has been on the local forces generated by muscle activation. As a result, there has been a significant amount of research on trying to understand muscles, recruitment, and activation. We have determined that a more direct approach is to deal with the joint space, i.e., the torque generated by these muscles on the joint (regardless of which muscle is active and which is recruited). Thus, our focus is the resultant action of these muscles on the joint. Because we use the DH parameterization method at each individual DOF, this approach lends itself well to our goals of estimating and predicting how the body is set in motion.

To set up the optimization problem, we will also need the gradients for all objective functions and constraints. Working in the joint space provides a direct and feasible method for accomplishing an effective optimization problem formulation.

As stated earlier, the most important benefit of this optimization formulation is that the equations of motion are not explicitly integrated, but are evaluated by inverse dynamics. The dynamic effort (performance measure) that is represented as the integral of the squares of all the joint torques is minimized. Indeed, dynamic balance is achieved by satisfying the ZMP constraint throughout the walking motion. Subsequently, the sequential quadratic programming algorithm is used to solve the nonlinear optimization problem. The results of the optimization problem, torque and joint profiles, are shown to be realistic when compared with the experimental data for normal walking. Besides normal walking, three other cases of walking with a shoulder backpack are simulated.

The objective of formulating the gait using predictive dynamics is to enable the prediction of natural motions, and to be able to simulate cause and effect, whereby a user can input various parameters, observe the simulated motion, and calculate the parameters of the motion (forces, torques, motion profiles, ground reaction forces [GRFs], and balance issues).

Because of the use of optimization to model the behavior and the biomechanics, it is believed that the algorithm drives the motion towards a more naturalistic and higher-fidelity motion. As a result, our objective is to obtain a high-fidelity motion with a model of a large DOF human skeleton. Because we use the joint space (not muscle space), we shall use the strength limits (strength surfaces) as the limits of what a person can do.

7.4 Spatial kinematics model

As detailed in Chapter 2, we shall use the DH parameterization method to model the full kinematics of the human body. Recall that the DH transformation matrix includes rotation and translation and is a function of four parameters, θ_i, d_i, α_i, and a_i, which relate coordinate frames i and i-1, depicted in Figure 7.1.

7.4.1 A kinematic 55-DOF human model

The spatial kinematic skeletal model with 55 DOFs (the z's), shown in Figure 7.2, will be used throughout this chapter to illustrate the principles. The model consists of six physical branches and one virtual branch. The physical branches are: the right leg, the left leg, the spine, the right arm, the left arm, and the head. In these branches, the right leg, the left leg and the spine start from the pelvis (z_4, z_5, z_6), while the right arm, left arm, and head start from the spine end joint (z_{30}, z_{31}, z_{32}).

The spine model includes four joints, and each joint has three rotational DOFs ($[z_{21}, z_{22}, z_{23}]$, $[z_{24}, z_{25}, z_{26}]$, $[z_{27}, z_{28}, z_{29}]$, $[z_{30}, z_{31}, z_{32}]$). The legs and arms are assumed to be symmetric with respect to the sagittal plane y-z. Each leg consists of a thigh, a shank, a rear foot, and a forefoot. There are seven DOFs for each leg: three at the hip joint (z_7, z_8, z_9), one at the knee joint (z_{10}), two at the ankle

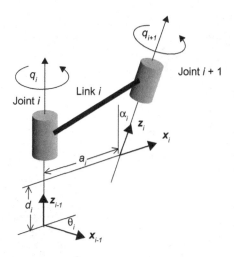

FIGURE 7.1

Joint coordinate systems using the Denavit–Hartenberg representation methodology.

joint (z_{11}, z_{12}), and finally, one to characterize the forefoot (z_{13}). At the clavicle, there are two orthogonal revolute joints (z_{33}, z_{34}). Each arm consists of an upper arm, a lower arm, and a hand. There are seven DOFs for each arm: three at the shoulder, two at the elbow, and two at the wrist. In addition, there are five DOFs for the head branch: three at the lower neck and two at the upper neck. The anthropometric data for the skeletal model representing a 50th percentile male, generated using GEBOD software (Cheng et al., 1994), are shown in Table 7.1. Note that these dimensions represent a specific human model—it is not a percentile-based anthropometry.

This 55-DOF skeletal model has been developed to simulate many human activities, including symmetric and asymmetric walking, running, climbing stairs, lifting objects, throwing, and many other tasks. For simulating each of these activities, some DOFs that do not participate in the activity in a significant manner are frozen to their neutral angles. The formulation presented here is quite flexible, allowing any DOF to be frozen to a specified value. In addition, limits on the range of motion of any DOF can be imposed. A general-purpose software has been developed that can be used to simulate these activities using the same skeletal model. For the symmetric gait simulation problem, the following DOFs are frozen: wrist joint, clavicle joint, neck joint, and two spine joints (shown as dashed enclosures in Figure 7.2). Therefore, the skeletal model used for gait simulation has 38 active DOFs. The other way to accomplish this objective would be to redefine the skeletal model for each activity. However, this would require redefinition of the body segments and recalculation of their mass and inertial properties, which would be quite tedious.

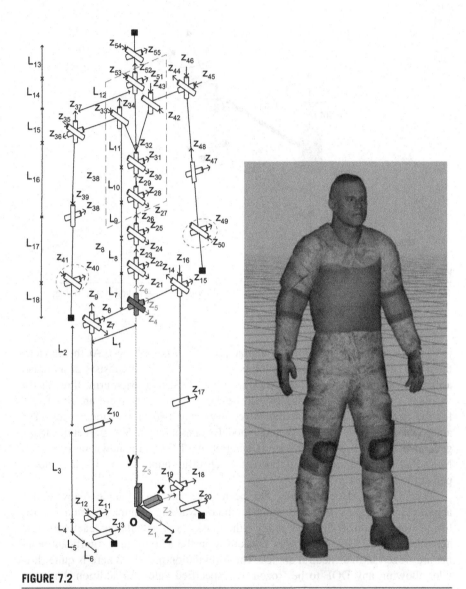

FIGURE 7.2

The 55-DOF digital human model (with global DOFs z_1, z_2, z_3, z_4, z_5, z_6).

7.4.2 Global DOFs and virtual joints

Consider six global DOFs that generate rigid body motion for the entire spatial skeleton model. The three translations are represented by three prismatic joints and the three rotations by three revolution joints in the DH method. We denote these joints as virtual joints to distinguish them from the physical human joints.

Table 7.1 Link Length and Mass Properties

Link	Length (cm)	Mass (kg)
L_1	8.51	4.48
L_2	38.26	9.54
L_3	39.46	3.74
L_4	5.0	0.5
L_5	9.01	0.7
L_6	7.56	0.23
L_7	9.0	2.32
L_8	5.63	2.32
L_9	5.44	2.32
L_{10}	6.0	2.32
L_{11}	17.39	3.0
L_{12}	16.76	5.78
L_{13}	20.0	4.22
L_{14}	17.1	1.03
L_{15}	4.41	2.8
L_{16}	25.86	1.9
L_{17}	24.74	1.34
L_{18}	16.51	0.5

We shall use these virtual joints to enable generalized global motion of the human with respect to a fixed coordinate frame at (0, 0, 0).

The two adjacent virtual joints are connected by a virtual link that uses zero mass and zero inertia to define the link properties. Finally, the virtual joints and links constitute a virtual branch that contains six global DOFs denoted by (z_1, z_2, z_3, z_4, z_5, z_6).

The virtual joints defined in the virtual branch not only generate global rigid body movements but also contain global generalized forces. These forces correspond to the six global DOFs: three forces (τ_1, τ_2, τ_3) and three moments (τ_4, τ_5, τ_6). For the system in dynamic equilibrium, these global generalized forces should be zero.

7.4.3 Forward recursive kinematics

We shall use the recursive kinematics and the Lagrangian approaches to carry out the kinematics and dynamic analyses of the 3D human model. The computational cost of the recursive formulation is $O(n)$. The sensitivity information about a joint involves only two adjacent joints; therefore, its computational cost is reduced to $O(n)$.

We use 4×4 matrices denoted by $\mathbf{A}_j$, $\mathbf{B}_j$, and $\mathbf{C}_j$ to represent recursive position, velocity, and acceleration transformation matrices for the jth joint, respectively. Given the link transformation matrix $(\mathbf{T}_j)$ and the kinematics state variables for each joint, angular displacement, velocity and acceleration (q_j, $\dot{q}_j$, and $\ddot{q}_j$), we have for $j = 1$ to n:

$$\mathbf{A}_j = \mathbf{T}_1\mathbf{T}_2\mathbf{T}_3 \cdots \mathbf{T}_j = \mathbf{A}_{j-1}\mathbf{T}_j \tag{7.1}$$

$$\mathbf{B}_j = \dot{\mathbf{A}}_j = \mathbf{B}_{j-1}\mathbf{T}_j + \mathbf{A}_{j-1}\frac{\partial \mathbf{T}_j}{\partial q_j}\dot{q}_j \tag{7.2}$$

$$\mathbf{C}_j = \dot{\mathbf{B}}_j = \ddot{\mathbf{A}}_j = \mathbf{C}_{j-1}\mathbf{T}_j + 2\mathbf{B}_{j-1}\frac{\partial \mathbf{T}_j}{\partial q_j}\dot{q}_j + \mathbf{A}_{j-1}\frac{\partial^2 \mathbf{T}_j}{\partial q_j^2}\dot{q}_j^2 + \mathbf{A}_{j-1}\frac{\partial \mathbf{T}_j}{\partial q_j}\ddot{q}_j \tag{7.3}$$

where $\mathbf{A}_0 = \mathbf{1}$ and $\mathbf{B}_0 = \mathbf{C}_0 = \mathbf{0}$. Then, the global position, velocity, and acceleration of a point in the Cartesian coordinate system can be calculated using the following formulas:

$$^0\mathbf{r}_j = \mathbf{A}_j\mathbf{r}_j; \quad ^0\dot{\mathbf{r}}_j = \mathbf{B}_j\mathbf{r}_j; \quad ^0\ddot{\mathbf{r}}_j = \mathbf{C}_j\mathbf{r}_j \tag{7.4}$$

where $^0\mathbf{r}_j$ and $\mathbf{r}_j$ are global and local augmented coordinates, respectively.

7.5 Dynamics formulation

7.5.1 Backward recursive dynamics

Based on forward recursive kinematics, the backward recursion for the dynamic analysis is accomplished by defining a 4×4 transformation matrix $\mathbf{D}_i$ and 4×1 transformation vectors $\mathbf{E}_i$, $\mathbf{F}_i$, and $\mathbf{G}_i$ as follows. Given the mass and inertia properties of each link, the external force $\mathbf{f}_k^T = \begin{bmatrix} ^kf_x & ^kf_y & ^kf_z & 0 \end{bmatrix}$ and the moment $\mathbf{h}_k^T = \begin{bmatrix} ^kh_x & ^kh_y & ^kh_z & 0 \end{bmatrix}$ for the link k, defined in the global coordinate system, the joint actuation torques derived in Chapter 4 (Equation 4.33) are τ_i for $i = n$ to 1 as:

$$\tau_i = tr\left[\frac{\partial \mathbf{A}_i}{\partial q_i}\mathbf{D}_i\right] - \mathbf{g}^T\frac{\partial \mathbf{A}_i}{\partial q_i}\mathbf{E}_i - \mathbf{f}_k^T\frac{\partial \mathbf{A}_i}{\partial q_i}\mathbf{F}_i - \mathbf{G}_i^T\mathbf{A}_{i-1}\mathbf{z}_0 \tag{7.5}$$

$$\mathbf{D}_i = \mathbf{I}_i\mathbf{C}_i^T + \mathbf{T}_{i+1}\mathbf{D}_{i+1} \tag{7.6}$$

$$\mathbf{E}_i = m_i^i\mathbf{r}_i + \mathbf{T}_{i+1}\mathbf{E}_{i+1} \tag{7.7}$$

$$\mathbf{F}_i = {}^k\mathbf{r}_f\delta_{ik} + \mathbf{T}_{i+1}\mathbf{F}_{i+1} \tag{7.8}$$

$$\mathbf{G}_i = \mathbf{h}_k\delta_{ik} + \mathbf{G}_{i+1} \tag{7.9}$$

where $\mathbf{D}_{n+1} = \mathbf{0}$ and $\mathbf{E}_{n+1} = \mathbf{F}_{n+1} = \mathbf{G}_{n+1} = \mathbf{0}$; $\mathbf{I}_i$ is the inertia matrix for link i; m_i is the mass of link i; $\mathbf{g}$ is the gravity vector; $^i\mathbf{r}_i$ is the location of center of mass of link i in the local frame i; $^k\mathbf{r}_f$ is the position of the external force in the local frame k; $\mathbf{z}_0 = \begin{bmatrix} 0 & 0 & 1 & 0 \end{bmatrix}^T$ for a revolute joint and $\mathbf{z}_0 = \begin{bmatrix} 0 & 0 & 0 & 0 \end{bmatrix}^T$ for a prismatic joint; and finally, δ_{ik} is Kronecker delta.

The first term in the torque expression (equation of motion) is the inertia and Coriolis torque; the second term is the torque due to gravity load; the third term is the torque due to external force; and the fourth term represents the torque due to the external moment.

7.5.2 Sensitivity analysis

In order to calculate the sensitivity, we must first find expressions for the derivatives of various quantities. The derivatives, $\dfrac{\partial \tau_i}{\partial q_k}, \dfrac{\partial \tau_i}{\partial \dot{q}_k}, \dfrac{\partial \tau_i}{\partial \ddot{q}_k}$ ($i = 1$ to n; $k = 1$ to n), can be evaluated for the articulated spatial human mechanical system in a recursive way using the foregoing recursive Lagrangian dynamics formulation:

$$
\frac{\partial \tau_i}{\partial q_k} = \begin{cases} tr\left(\dfrac{\partial^2 \mathbf{A}_i}{\partial q_i \partial q_k}\mathbf{D}_i + \dfrac{\partial \mathbf{A}_i}{\partial q_i}\dfrac{\partial \mathbf{D}_i}{\partial q_k}\right) - \mathbf{g}^T \dfrac{\partial^2 \mathbf{A}_i}{\partial q_i \partial q_k}\mathbf{E}_i - \mathbf{f}^T \dfrac{\partial^2 \mathbf{A}_i}{\partial q_i \partial q_k}\mathbf{F}_i - \mathbf{G}_i^T \dfrac{\partial \mathbf{A}_{i-1}}{\partial q_k}\mathbf{z}_0 & (k \le i) \\[4mm] tr\left(\dfrac{\partial \mathbf{A}_i}{\partial q_i}\dfrac{\partial \mathbf{D}_i}{\partial q_k}\right) - \mathbf{g}^T \dfrac{\partial \mathbf{A}_i}{\partial q_i}\dfrac{\partial \mathbf{E}_i}{\partial q_k} - \mathbf{f}^T \dfrac{\partial \mathbf{A}_i}{\partial q_i}\dfrac{\partial \mathbf{F}_i}{\partial q_k} & (k > i) \end{cases}
$$

$$(7.10)$$

$$\frac{\partial \tau_i}{\partial \dot{q}_k} = tr\left(\frac{\partial \mathbf{A}_i}{\partial q_i}\frac{\partial \mathbf{D}_i}{\partial \dot{q}_k}\right) \tag{7.11}$$

$$\frac{\partial \tau_i}{\partial \ddot{q}_k} = tr\left(\frac{\partial \mathbf{A}_i}{\partial q_i}\frac{\partial \mathbf{D}_i}{\partial \ddot{q}_k}\right) \tag{7.12}$$

A more detailed treatment on the derivation of sensitivity equations can be found in some of the published work on sensitivity (Xiang et al., 2008).

7.5.3 Mass and inertia property

Predictive dynamics uses forward kinematics to transfer the motion from the origin towards the end-effector along the branch as shown in Figure 7.3. This process involves only state variables and geometrical parameters. However, backward dynamics propagates forces from the end-effector to the origin. Mass and inertia property of the links are taken into consideration for dynamic analysis.

In Figure 7.3, joint (k) and joint ($k + 1$) are connected by link ($j + 1$) for which mass and inertia properties are defined in the local coordinate z_{i+3}. The links between coordinates z_{i+3} and z_{i+2}, and z_{i+2} and z_{i+1} have zero link length, and zero mass and inertia, so that the force is correctly transferred back through z_{i+3}, z_{i+2}, and z_{i+1} for the joint (k).

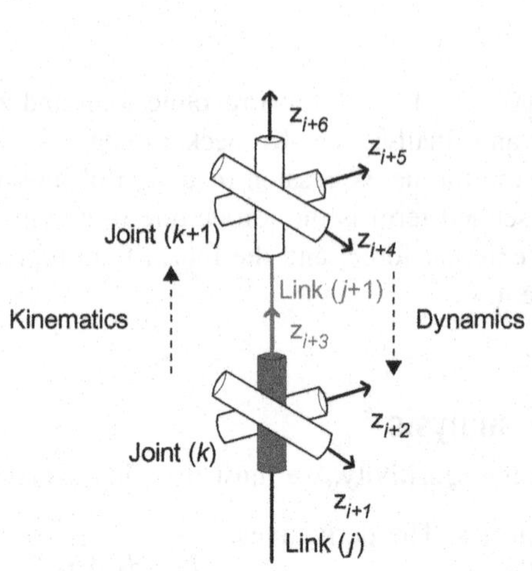

FIGURE 7.3

Mass and inertia allocation for joint pairs.

7.6 Gait model

7.6.1 One-step gait model

We shall present a modeling method for a complete gait cycle that includes two continuous steps (also called one stride). In the present formulation, normal walking is assumed to be symmetric and cyclic; therefore only one step of the gait cycle needs to be modeled and simulated. Each step is divided into two phases: single support phase and double support phase.

The single support phase occurs when one foot contacts the ground while the other leg is swinging; it starts from the rear foot toe-off and ends when the swinging foot lands on the ground with a heel strike; the time duration for this phase is denoted as T_{SS}.

Considering the ball joint of foot, the single support phase can be divided into two basic supporting modes: rear foot single support and forefoot single support. The double support phase is characterized by both feet contacting the ground. This phase starts from the front foot heel strike and ends with the contra-lateral foot toe-off. The time duration of double support is denoted as T_{DS}. In this work, a walking step starts from the left heel strike, then goes through left foot flat, right toe-off, right leg swing, left heel off, and finally comes back to right heel strike as shown in Figure 7.4. The foot support polygon is plotted in Figure 7.5. The foot contacting conditions are summarized in Table 7.2.

Symmetry conditions for the gait cycle are needed so that only one step can be modeled to simulate gait. The successive step repeats the motion of the previous step by swapping the roles of the legs and arms. The initial and final joint angles and velocities (at left heel strike and subsequent right heel strike) should satisfy symmetry conditions so as to generate continuous and cyclic gait motion.

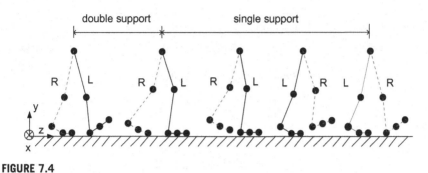

FIGURE 7.4

Basic feet supporting modes in a step (side view: R denotes right leg; L denotes left leg).

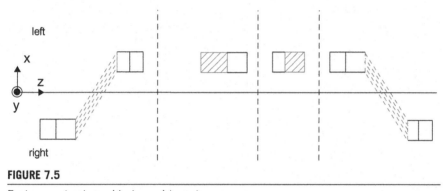

FIGURE 7.5

Foot support polygon (dash area) in a step.

7.6.2 Ground reaction forces (GRF)

It is necessary to calculate GRFs in the gait formulation for multiple purposes. Ground reaction forces provide rigorous validation criteria for whether the walk prediction is realistic and whether it does indeed simulate a natural human walk. To include the GRF in the formulation, we have two challenges: not only is the GRF value transient but also the GRF position is variable. To solve for the GRF, a two-step algorithm is developed.

Initially we will distinguish forces into two categories: active forces and GRF. Active forces include inertia, Coriolis, gravity, and external forces and moments. The main idea of the algorithm is to first calculate the resultant of the active forces and ZMP location from the equations of motion, and then calculate GRF using the global equilibrium conditions between the active forces and GRF. After that, the obtained GRF are applied as external loads at the ZMP, together with the

Table 7.2 Foot Contacting Conditions: Four Modes in a Step

Double support	Single support		Double support
	Rear-foot	**Fore-foot**	
Right toe	Left heel	Left ball	Left toe
Left heel	Left ball	Left toe	Right heel

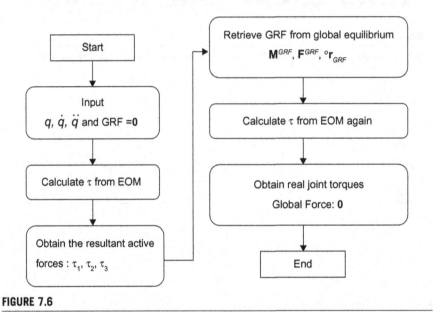

FIGURE 7.6

Flow chart of the two-step algorithm to obtain GRF and real joint torques.

active forces, to recover the real joint torques. This two-step algorithm is depicted in Figure 7.6 and explained as follows:

>*Step 1*: Given current state variables q, $\dot{q}$, and $\ddot{q}$, external loads and gravity, the joint torques are calculated using the inverse recursive Lagrangian dynamics without the GRF. This is illustrated in Figure 7.6. The global forces τ_1, τ_2, τ_3 and moments τ_4, τ_5, τ_6 in the virtual branch are not zero at this stage due to exclusion of GRF. These forces are in fact the resultants of the active forces at the end of the virtual branch, i.e., the pelvis. After that, the ZMP is calculated using these forces, as explained in the following subsection.
>
>*Step 2*: This step is illustrated in Figure 7.6. Considering the global equilibrium between the resultant active forces and the GRF at ZMP, the resultant GRF are obtained and then treated as external forces applied at ZMP for the human model, such as the forces $\mathbf{f}$ and $\mathbf{h}$ in Equations (7.5−7.9). Given the state

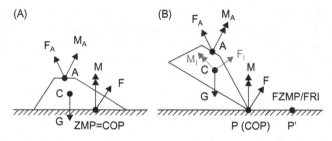

FIGURE 7.7

Balanced and unbalanced states: (A) balanced state, (B) unbalanced state.

variables, external loads, and gravity loads as well as the GRF, the real joint torques are recovered from the equations of motion using Equation (7.5).

7.7 Zero-Moment point (ZMP)

The ZMP approach is a well-established bipedal dynamic balance criterion that has been widely used in the fields of robotics and biomechanics (Vukobratović and Borovac, 2004). The ZMP is the point on the ground where the resultant tangential moments of the active forces are zero, as illustrated in Figure 7.7.

Here, the forces are distinguished into two categories: active forces and passive forces. Active forces include inertia, Coriolis, gravity, and external forces. Passive forces are the GRFs. The balanced and unbalanced states of the human system are illustrated in Figure 7.7, where point A denotes the ankle, point C is the center of mass of foot, $\mathbf{G}$ is foot gravity force, $\mathbf{F_A}$ and $\mathbf{M_A}$ are resultant force and moment of body parts excluding the contacting foot, $\mathbf{F}$ and $\mathbf{M}$ are the resultant GRFs, and $\mathbf{F_I}$ and $\mathbf{M_I}$ are the inertia forces of the foot due to rotation. FZMP represents fictitious ZMP, and FRI is foot rotation indicator.

The position of ZMP can be calculated using the conditions: $M_z = 0$ and $M_x = 0$ (z is the walking direction, x is the lateral direction, and y is the vertical direction; refer to Figure 7.4).

$$x_{zmp} = \frac{\sum_{i=1}^{nlink}(m_i(-\ddot{y}_i + g)x_i + m_i\ddot{x}_i y_i - J_i\ddot{\theta}_{iz} + f_{iy}x_i - f_{ix}y_i + h_{iz})}{\sum_{i=1}^{nlink} m_i(-\ddot{y}_i + g)} \qquad (7.13a)$$

$$z_{zmp} = \frac{\sum_{i=1}^{nlink}(m_i(-\ddot{y}_i + g)z_i + m_i\ddot{z}_i y_i + J_i\ddot{\theta}_{ix} + f_{iy}z_i - f_{iz}y_i - h_{ix})}{\sum_{i=1}^{nlink} m_i(-\ddot{y}_i + g)} \qquad (7.13b)$$

where x_i, y_i, z_i are the global coordinates of the center of mass for the link i; m_i is the mass, J_i is the global inertia, $\ddot{\theta}_i$ is the global angular acceleration with link i; f_i and h_i are the external force and moment applied on link i; and $g = -9.8062 \text{ m/s}^2$.

To calculate ZMP, the inertia J_i and angular acceleration $\ddot{\theta}_i$ must be evaluated in the global coordinates; however, they have been defined in the local coordinates associated to link i in the DH method. The transformation to the global coordinates is tedious and time-consuming. Therefore, some researchers simply ignore these terms in ZMP calculation.

To overcome this difficulty, an alternative method is developed to calculate ZMP based on global equilibrium condition, or equivalently, the forces and moments in the virtual branch obtained from the equations of motion. The basic idea is to use the resultant of the active forces and moments to calculate ZMP directly instead of evaluating them link by link as in Equations (7.13).

Given state variables (q_j, $\dot{q}_j$, and $\ddot{q}_j$) for each joint, apply active forces to the mechanical system, excluding GRF. The calculated generalized forces (τ_1, τ_2, τ_3, τ_4, τ_5, τ_6) in the virtual branch from the equations of motion are in fact the resultant active forces and moments. After obtaining the resultant of the active forces, we can use them to calculate ZMP using the following three steps:

1. Calculate the resultant of the active forces and moments at the pelvis in the inertial reference frame;
2. Transfer these forces and moments to the origin of the inertial reference frame (o-xyz);
3. Calculate the ZMP from its definition.

We present more details of these three steps below.

7.7.1 Global forces at the pelvis

We denote the forces at the pelvis by $\mathbf{F}^p = [\,F_x^p \quad F_y^p \quad F_z^p\,]^T$.

The direction of the resultant active moments (τ_5, τ_6) at the pelvis are defined in the local coordinates (z_5, z_6). Because of the global rotational movements (q_4, q_5), the forces at the pelvis no longer align with the global Cartesian coordinates (o-xyz), as shown in Figure 7.8. Since ZMP is defined in the global Cartesian coordinates, we need to recover the resultant active moments $\mathbf{M}^p = [\,M_x^p \quad M_y^p \quad M_z^p\,]^T$ at the pelvis in the global Cartesian coordinates. This is accomplished using the following equilibrium equation:

$$
\begin{bmatrix}
\cos(z_4,x) & \cos(z_4,y) & \cos(z_4,z) \\
\cos(z_5,x) & \cos(z_5,y) & \cos(z_5,z) \\
\cos(z_6,x) & \cos(z_6,y) & \cos(z_6,z)
\end{bmatrix}
\begin{bmatrix} M_z^p \\ M_x^p \\ M_y^p \end{bmatrix}
+
\begin{bmatrix} \tau_4 \\ \tau_5 \\ \tau_6 \end{bmatrix}
= \mathbf{0}
\qquad (7.14)
$$

where τ_4, τ_5, τ_6 are resultant moments along the DH local axes associated with their DOFs; $\cos(z_4,x) = 0$, $\cos(z_4,y) = 0$, and $\cos(z_4,z) = 1$ because the first rotational joint is aligned with global z-axis.

The resultant active forces $\mathbf{F}^p = [\,F_x^p \quad F_y^p \quad F_z^p\,]^T$ at the pelvis are obtained by considering the equilibrium between two sets of forces as follows:

$$
F_x^p + \tau_2 = 0; \quad F_y^p + \tau_3 = 0; \quad F_z^p + \tau_1 = 0
\qquad (7.15)
$$

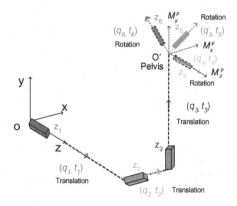

FIGURE 7.8

Global DOFs in virtual branch.

7.7.2 Global forces at origin

After obtaining the global forces at the pelvis in the global Cartesian coordinates we can transfer the resultant active force from the pelvis to the origin using the equilibrium conditions. Thus, the resultant active forces ($\mathbf{M}^o$, $\mathbf{F}^o$) at the origin are obtained as follows:

$$\mathbf{M}^o = \mathbf{M}^p + {}^o\mathbf{r}_p \times \mathbf{F}^p$$
$$\mathbf{F}^o = \mathbf{F}^p \tag{7.16}$$

where $\mathbf{M}^o = [\, M_x^o \quad M_y^o \quad M_z^o \,]^T$ and $\mathbf{F}^o = [\, F_x^o \quad F_y^o \quad F_z^o \,]^T$; ${}^o\mathbf{r}_p$ is the pelvis position vector in the global coordinate system, as depicted in Figure 7.9.

7.7.3 ZMP calculation

Next, the resultant active forces are further transferred from the origin to the ZMP by using the equilibrium conditions; i.e., ($\mathbf{M}^{zmp}$, $\mathbf{F}^{zmp}$) are obtained using the equation:

$$\begin{pmatrix} M_x^{zmp} \\ M_y^{zmp} \\ M_z^{zmp} \end{pmatrix} = \begin{pmatrix} M_x^o \\ M_y^o \\ M_z^o \end{pmatrix} + \begin{pmatrix} x_{zmp} \\ y_{zmp} \\ z_{zmp} \end{pmatrix} \times \begin{pmatrix} F_x^o \\ F_y^o \\ F_z^o \end{pmatrix} \tag{7.17}$$
$$\mathbf{F}^{zmp} = \mathbf{F}^o$$

where ${}^o\mathbf{r}_{zmp} = [\, x_{zmp} \quad y_{zmp} \quad z_{zmp} \,]^T$ is the ZMP position vector in the global coordinates. Since ZMP is set on the level ground and tangential moments are zero due to its definition, we have:

$$y_{zmp} = 0; \ M_x^{zmp} = 0; \ M_z^{zmp} = 0 \tag{7.18}$$

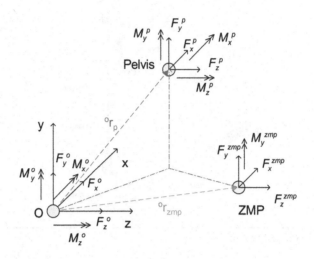

FIGURE 7.9

Resultant active forces at the pelvis, origin and ZMP.

Using Equations (7.17) and (7.18), the ZMP position is uniquely obtained as follows:

$$x_{zmp} = \frac{M_z^o}{F_y^o}; \quad z_{zmp} = \frac{-M_x^o}{F_y^o} \tag{7.19}$$

In addition, the resultant active moment at ZMP along the y-axis is also obtained from Equation (7.17):

$$M_y^{zmp} = M_y^o + F_x^o z_{zmp} - F_z^o x_{zmp} \tag{7.20}$$

There are two major advantages of using the foregoing ZMP formulation: one is that the calculation of resultant active forces from the equations of motion is very convenient and straightforward; the other is that the resultant active forces ($\mathbf{M}^{zmp}$, $\mathbf{F}^{zmp}$) and position $^o\mathbf{r}_{zmp}$ of the ZMP are obtained simultaneously. The resultant active forces at ZMP are used to calculate the GRFs.

7.8 Calculating ground reaction forces (GRF)

The resultant GRF are located at the center of pressure (COP), which coincides with the ZMP as long as the dynamic system is in balance (Goswami, 1999; Sardain and Bessonnet, 2004). Thus, the transient position of the resultant GRF can be obtained by tracing the ZMP position $^o\mathbf{r}_{zmp}$ using Equation (7.19).

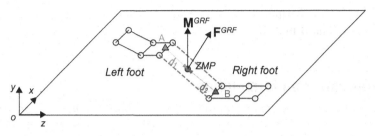

FIGURE 7.10

Partition of ground reaction forces.

Moreover, the transient value of resultant GRF is also obtained from global equilibrium conditions as:

$$\mathbf{M}^{GRF} + \mathbf{M}^{zmp} = 0$$
$$\mathbf{F}^{GRF} + \mathbf{F}^{zmp} = 0 \qquad (7.21)$$
$$^{o}\mathbf{r}_{GRF} - {}^{o}\mathbf{r}_{zmp} = 0$$

In the single support phase, one foot supports the whole body and the ZMP stays in the foot area so that GRF can be applied at the ZMP directly. However, in the double support phase, the ZMP is located between the two supporting feet, and the resultant GRF needs to be distributed to the two feet appropriately. This partitioning process can be treated as a sub-optimization problem (Dasgupta and Nakamura, 1999). In order to simplify this process, the GRF is distributed to the points (A, B) of the supporting parts on each foot as shown in Figure 7.10, where point A (triangle) is the left toe center and point B (triangle) is the right heel center; d_1 and d_2 are the distances from the ZMP (circle) to points A and B respectively. A linear relationship is used to partition GRF. The GRF value is first linearly decomposed at the ZMP as follows:

$$\mathbf{M}_1^{GRF} = \frac{d_2}{d_1 + d_2}\mathbf{M}^{GRF}, \quad \mathbf{F}_1^{GRF} = \frac{d_2}{d_1 + d_2}\mathbf{F}^{GRF}$$
$$\mathbf{M}_2^{GRF} = \frac{d_1}{d_1 + d_2}\mathbf{M}^{GRF}, \quad \mathbf{F}_2^{GRF} = \frac{d_1}{d_1 + d_2}\mathbf{F}^{GRF} \qquad (7.22)$$

Then, $(\mathbf{M}_1^{GRF}, \mathbf{F}_1^{GRF})$ are transferred to point A and $(\mathbf{M}_2^{GRF}, \mathbf{F}_2^{GRF})$ to point B as follows:

$$\mathbf{M}^A = \mathbf{M}_1^{GRF} + {}^{A}\mathbf{d}_{zmp} \times \mathbf{F}_1^{GRF}$$
$$\mathbf{F}^A = \mathbf{F}_1^{GRF} \qquad (7.23)$$

$$\mathbf{M}^B = \mathbf{M}_2^{GRF} + {}^{B}\mathbf{d}_{zmp} \times \mathbf{F}_2^{GRF}$$
$$\mathbf{F}^B = \mathbf{F}_2^{GRF} \qquad (7.24)$$

where $^A\mathbf{d}_{zmp}$ is the position vector from point A to ZMP, and $^B\mathbf{d}_{zmp}$ is the position vector from point B to ZMP.

7.9 Optimization formulation

The objective of this section is to formulate the optimization problem for walking in terms of three major components: design variables, objective function, and constraints. At a given walking velocity (V) and a step length (L), the time duration for the step is calculated as $T = L/V$. The double support time duration is taken as $T_{DS} = \alpha T$; as a result, the single support time duration is given as $T_{SS} = (1 - \alpha)T$. Single support is detailed into rear foot support (mid-stance) and forefoot support (terminal stance). Their time durations are set to βT_{SS} and $(1 - \beta)T_{SS}$, respectively. The parameters α and β are obtained from the literature (Ayyappa, 1997).

The walking task is formulated as a nonlinear optimization problem. A general mathematical form is defined as: Find the optimal joint trajectories $\mathbf{q}(t)$ and joint torques $\boldsymbol{\tau}(t)$ to minimize a human performance measure subject to physical constraints:

$$
\begin{aligned}
&\text{Find}: \quad \mathbf{q},\ \boldsymbol{\tau} \\
&\text{To}: \quad \min F(\mathbf{q},\ \boldsymbol{\tau}) \\
&\text{Sub}. \quad h_i = 0;\ i = 1,\ldots,m \\
&\qquad\quad g_j \leq 0;\ j = 1,\ldots,k
\end{aligned}
\tag{7.25}
$$

where h_i are the equality constraints and g_j are the inequality constraints.

7.9.1 Design variables

In the current formulation, the design variables are the joint angle profiles $\mathbf{q}(t)$. The joint torques $\boldsymbol{\tau}(\mathbf{q}, t)$ are calculated using the governing differential equations. This is called the inverse dynamics procedure where the differential equations are not integrated. This has also been called the differential inclusion formulation.

7.9.2 Objective function

The predicted motion depends strongly on the adopted objective function F. In this work, the dynamic effort, the time integral of squares of all the joint torques, is used as the performance criterion for the walking problem:

$$
F(\mathbf{q}) = \int_{t=0}^{T} \left(\frac{\boldsymbol{\tau}(\mathbf{q}, t)}{|\tau|_{\max}} \right)^{T} \cdot \left(\frac{\boldsymbol{\tau}(\mathbf{q}, t)}{|\tau|_{\max}} \right) dt
\tag{7.26}
$$

where $|\tau|_{\max}$ is the maximum absolute value of joint torque limit.

Figure 7.11 illustrates the optimization problem containing the three components of a formulation. Note that we have used the simple form of the minimization function as the integration of the torque square versus the normalized form in Equation (7.26).

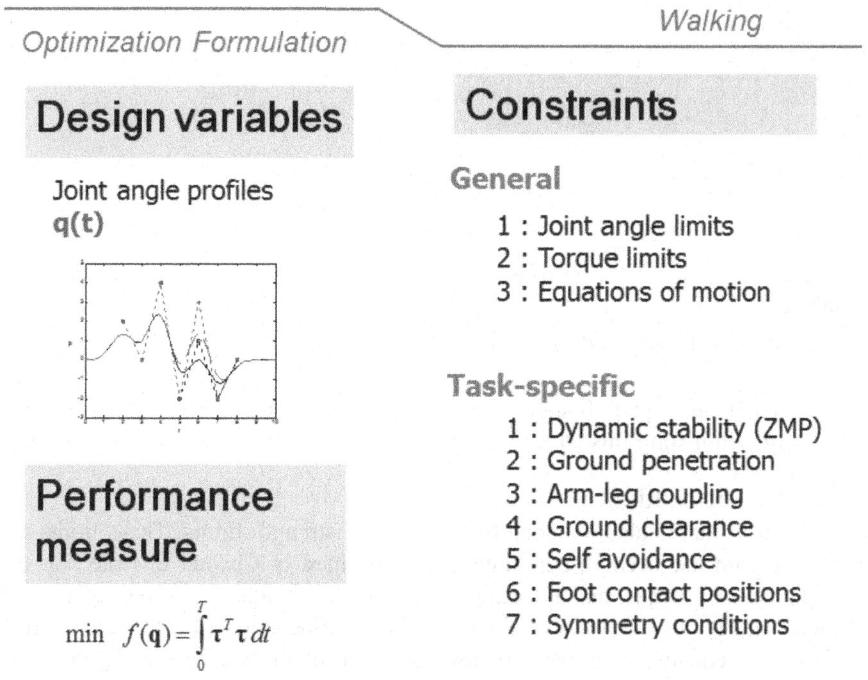

Optimization Formulation   *Walking*

Design variables

Joint angle profiles
q(t)

Constraints

General

 1 : Joint angle limits
 2 : Torque limits
 3 : Equations of motion

Task-specific

 1 : Dynamic stability (ZMP)
 2 : Ground penetration
 3 : Arm-leg coupling
 4 : Ground clearance
 5 : Self avoidance
 6 : Foot contact positions
 7 : Symmetry conditions

Performance measure

$$\min \ f(\mathbf{q}) = \int_0^T \boldsymbol{\tau}^T \boldsymbol{\tau} \, dt$$

FIGURE 7.11

The general optimization formulation for walking.

7.9.3 Constraints

Two types of constraints are encountered for the walking optimization problem: one is the time-dependent constraints, which include joint limits, torque limits, ground penetration, dynamic balance, arm-leg coupling, and self-avoidance. These constraints are imposed throughout the time interval. The second type is the time-independent constraints, which comprise the symmetry conditions, ground clearance, and initial and final foot positions; these constraints are considered only at a specific time point during the step.

7.9.3.1 Time-dependent constraints
7.9.3.1.1 Joint limits

To avoid hyperextension, the joint limits are taken into account in the formulation. The joint limits representing the physical range of motion are

$$\mathbf{q}^L \leq \mathbf{q}(t) \leq \mathbf{q}^U, \quad 0 \leq t \leq T \tag{7.27}$$

where $\mathbf{q}^L$ are the lower joint limits and $\mathbf{q}^U$ the upper limits as presented later in Table 7.3. A joint limit constraint is also used to "freeze" a DOF by setting its

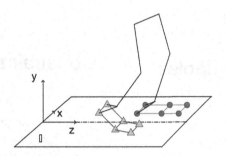

FIGURE 7.12

Foot ground penetration conditions.

lower bound and upper bound to the neutral angle (the natural angle at rest) instead of eliminating this DOF from the skeleton model.

7.9.3.1.2 Strength limits

Each joint torque is also bounded by its physical strength limits. These limits are obtained from the strength experiments as presented in Chapter 6. Note that the percent max of torque that is acting as a limit is a single point on the surface depicted in Figures 6.8 and 6.9. Note that these surfaces must be represented into a parametric equation in order to be used in the inequality constraint as follows:

$$\tau^L \leq \tau(t) \leq \tau^U, \quad 0 \leq t \leq T \tag{7.28}$$

where τ^L are the lower torque limits and τ^U the upper limits.

7.9.3.1.3 Ground penetration

Walking is characterized with unilateral contact between the foot and ground as shown in Figure 7.12. While the foot contacts the ground, the height and velocity of the contacting points (circles) are zero. In contrast, the height of other points (triangles) on the foot is greater than zero.

Therefore, the ground penetration constraints are formulated as follows:

$$\begin{aligned} y_i(t) &= 0, \quad \dot{x}_i(t) = 0, \quad \dot{y}_i(t) = 0, \quad \dot{z}_i(t) = 0, \quad i \in \Omega \\ y_i(t) &\geq \varepsilon, \quad\quad\quad i \notin \Omega, \quad\quad\quad 0 \leq t \leq T \end{aligned} \tag{7.29}$$

where ε is a small positive number and Ω is the set of contacting points.

7.9.3.1.4 Dynamic balance

The dynamic balance is achieved by forcing the ZMP to remain within the foot support polygon (FSP) as depicted in Figure 7.13, where Γ is a vector along the boundary of the FSP and $\mathbf{r}$ is the position vector from a vertex of the FSP to the ZMP.

The ZMP constraint is mathematically expressed as follows:

$$(\mathbf{r}_i \times \Gamma_i) \cdot \mathbf{n}_y \leq 0, \quad i = 1, \ldots, 4 \tag{7.30}$$

where $\mathbf{n}_y$ is the unit vector along the y-axis.

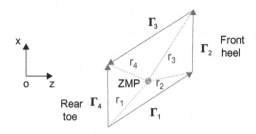

FIGURE 7.13

Foot support polygon (top view).

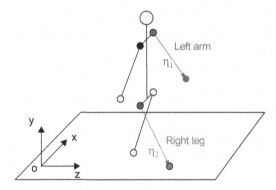

FIGURE 7.14

Arm-leg coupling motion.

7.9.3.1.5 Arm-leg coupling

It is believed that the arm-swing is performed to help balance the upper body during walking to reduce the trunk moment in the vertical direction. Swinging arms are not necessary or required for motion; however, that motion provides dynamic balance.

In practice, it is difficult to measure the moment produced by the swing arm. In this formulation, we introduce a two-pendulum model to represent arm-leg coupling kinematics during the walking motion. The basic idea of the arm-leg coupling constraint is that the arm-swing on one side counteracts the leg-swing on the other side as depicted in Figure 7.14, where the first pendulum η_1 represents the left arm (from the left shoulder to the left wrist), and the second pendulum η_2 denotes the right leg (from the right hip to the right ankle).

The mathematical form of a coupling constraint is written as:

$$(\eta_1 \cdot \mathbf{n}_z)(\eta_2 \cdot \mathbf{n}_z) \geq 0 \tag{7.31}$$

where $\mathbf{n}_z$ is the unit vector along the z-axis. It is important to note that the arm-leg coupling constraint is imposed only on the swing directions of the arm and

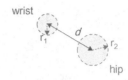

FIGURE 7.15

Self-avoidance constraint between the wrist and hip.

the leg rather than a quantitative relationship on the swing angles. The swing angles are determined in the optimization process.

7.9.3.1.6 Self-avoidance

Self-avoidance is considered in this formulation to prevent penetration of the arm in the body. A sphere-filling algorithm is used to formulate this constraint, as shown in Figure 7.15:

$$d(\mathbf{q}, t) - r_1 - r_2 \geq 0, \quad 0 \leq t \leq T \tag{7.32}$$

where r_1 is a constant radius to represent the wrist, r_2 is another radius to represent the hip; and d is the distance between the wrist and hip.

7.9.3.2 Time-independent constraints

The following time-independent constraints are imposed on the optimization problem.

7.9.3.2.1 Symmetry conditions

The gait simulation starts from the left heel strike and ends with the right heel strike. The initial and final postures and velocities should satisfy the symmetry conditions to generate continuous walking motion. These conditions are expressed as follows:

$$\begin{aligned}
q_L(0) - q_R(T) &= 0 & \dot{q}_L(0) - \dot{q}_R(T) &= 0 \\
q_{Sx}(0) - q_{Sx}(T) &= 0 & \dot{q}_{Sx}(0) - \dot{q}_{Sx}(T) &= 0 \\
q_{Sy}(0) + q_{Sy}(T) &= 0 & \dot{q}_{Sy}(0) + \dot{q}_{Sy}(T) &= 0 \\
q_{Sz}(0) + q_{Sz}(T) &= 0 & \dot{q}_{Sz}(0) + \dot{q}_{Sz}(T) &= 0
\end{aligned} \tag{7.33}$$

where subscripts L and R represent the DOFs of the leg, arm, and shoulder joints which satisfy the symmetry conditions with the contra-lateral leg, arm and shoulder joints; the subscript S represents the DOFs of the spine, neck and global joints which satisfy the symmetry conditions on themselves at the initial and final times; x, y, z are the global axes.

7.9.3.2.2 Ground clearance

To avoid foot drag motion, a ground clearance constraint is imposed during the walking motion. Instead of controlling the maximum height of the swing leg, the maximum knee flexion at mid-swing is used to formulate the ground clearance constraint. Biomechanical experiments have shown that the maximum knee flexion of normal gait is around 60 degrees regardless of the subject's age and gender. This constraint is expressed as

$$-\varepsilon \le q_{knee} - 60 \le \varepsilon, \quad t = t_{midswing} \tag{7.34}$$

where ε is a small range of motion, i.e., $\varepsilon = 5$ degrees.

7.9.3.2.3 Initial and final foot contacting position

Since the step length L is given, the foot initial and final contacting positions are specified at the initial and final times to satisfy the step length constraint. It is noted that the initial and final postures and velocities are determined by the optimization process instead of specified from the experiments.

$$\begin{aligned} \mathbf{x}_i(0) &= \tilde{\mathbf{x}}_i(0), \\ \mathbf{x}_i(T) &= \tilde{\mathbf{x}}_i(T), \quad i \in \Omega \end{aligned} \tag{7.35}$$

where $\tilde{\mathbf{x}}_i$ is the specified initial and final contacting position, and Ω is the set of contacting points as specified in Table 7.2.

7.10 Numerical discretization

In order for the numerical solver to calculate the design variables, the time domain is discretized by using cubic B-spline curves, which are defined by a set of control points $\mathbf{P}$ and time grid points (knots) $\mathbf{t}$. A joint profile $q(t)$ is parameterized by using B-splines as follows:

$$q(\mathbf{t}, \mathbf{P}) = \sum_{i=1}^{m} B_i(\mathbf{t}) p_i \quad 0 \le t \le T \tag{7.36}$$

where $B_i(\mathbf{t})$ are the basis functions, $\mathbf{t} = \{t_0, \ldots, t_s\}$ is the knot vector, and $\mathbf{P} = \{p_1, \ldots, p_m\}$ is the control points vector. With this representation, the control points become the optimization variables (also called the design variables).

Since q, $\dot{q}$, and $\ddot{q}$ are functions of $\mathbf{t}$ and $\mathbf{P}$, torque $\tau = \tau(\mathbf{t}, \mathbf{P})$ is an explicit function of the knot vector and control points. Thus, the derivatives of a torque with respect to the control points can be computed using the chain rule as

$$\frac{\partial \tau}{\partial P_i} = \frac{\partial \tau}{\partial q}\frac{\partial q}{\partial P_i} + \frac{\partial \tau}{\partial \dot{q}}\frac{\partial \dot{q}}{\partial P_i} + \frac{\partial \tau}{\partial \ddot{q}}\frac{\partial \ddot{q}}{\partial P_i} \tag{7.37}$$

Multiplicity is used in the knot vector at the end points of the time interval. This property guarantees that the starting and ending joint angle values of a DOF are exactly those corresponding to the first and last control point values. This makes it easier to impose the symmetric posture constraints. The time-dependent constraints are imposed not only at the knot points but also between the knots, so that a very smooth motion is generated.

7.11 Example: predicting the gait

The examples presented below will demonstrate predictive dynamics as a method for predicting normal walking, cause and effect, and symmetric and asymmetric gaits.

We have chosen to use a sequential quadratic programming (SQP) algorithm as implemented in SNOPT (Gill et al., 2002) to solve the optimization problem. To use the algorithm, cost and constraint functions and their gradients need to be calculated. The foregoing recursive kinematics and dynamics procedures provide accurate gradients to improve the computational efficiency of the optimization algorithm. Appropriate normal walking parameters (velocity and step length) are obtained from Inman et al. (1981). In addition to normal walking, the current work also considers situations where people walk and carry backpacks with various weights (20 lbs, 40 lbs, and 80 lbs).

7.11.1 Normal walking

Walking speeds in humans vary greatly depending on factors such as anthropometry, weight, age, terrain, surface, load, culture, effort, and fitness. The average human walking speed is about 5.0 km/h, or 1.4 m/s, about 3.1 mph. Walking research has found pedestrian walking speeds ranging from 4.51 km/h to 4.75 km/h for older individuals and 5.32 km/h to 5.43 km/h for younger individuals, although a brisk walking speed can be around 6.5 km/h and champion race walkers can average more than 14 km/h over a distance of 20 km.

For the problem defined herein, we shall assume a normal gait motion, where the user inputs normal walking velocity $V = 1.2$ m/s and step length $L = 0.6$ m. There are 330 design variables (55 DOFs, each with 6 control points) and 1036 nonlinear constraints. First the optimization problem is solved to obtain a feasible solution for the walking problem. Here $\mathbf{q} = \mathbf{0}$ is used as the starting point with $F(\mathbf{q}) = constant$ as the objective function and all constraints imposed. This is a new procedure for obtaining feasible solutions for a nonlinear programming problem that has proved to be very effective in testing feasibility of the predictive dynamics formulation.

Once a feasible solution has been obtained, it is used as the starting point for the optimization problem with dynamic effort as the objective function. There are two advantages of obtaining a feasible solution first: one is to test the feasibility of the problem formulation; and the second is to obtain a good starting point for the original optimization problem. The optimality and feasibility tolerances are both set to $\varepsilon = 10^{-3}$ for SNOPT and the optimal solution is obtained in 512 CPU seconds on a Pentium(R) 4, 3.46-GHz computer. There are 158 active constraints at the optimal solution.

Figure 7.16 shows the resulting stick diagram of a 3D human walking on level ground and includes the motion in the single support phase and the double support phase. As expected, correct knee bending occurs to avoid collision with the

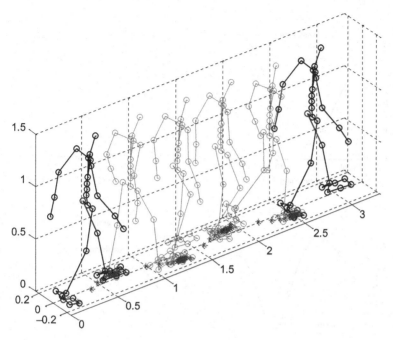

FIGURE 7.16

The diagram of optimized cyclic walking motion (two strides).

ground. The arms swing to balance the leg swing. The continuity condition is satisfied to generate smooth walking motion where the initial and final postures are also optimized. The ZMP location is also plotted in the figure and it stays in the foot support polygon to satisfy the dynamic balance condition. It is important to note that the spine keeps upright automatically to reduce energy expenditure of the walking motion.

7.11.1.1 Kinematics

There are six well-known kinematic variables (angles and displacements) that have been established as the determinants of forward walking (Saunders et al., 1953). These determinants correspond to the lower extremities and pelvic motion as follows:

a. Hip flexion/extension
b. Knee flexion/extension
c. Ankle plantar/dorsiflexion
d. Pelvic tilt
e. Pelvic rotation
f. Lateral pelvic displacement.

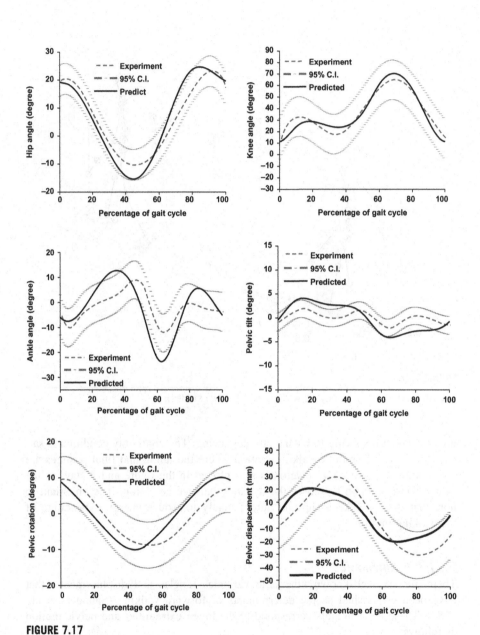

FIGURE 7.17

Comparison of predicted determinants with the experimental data.

The six determinants of normal walking predicted using predictive dynamics for a stride, starting from heel strike and ending with the same heel strike, are plotted in Figure 7.17. A more detailed treatment of validation methods and results for tasks predicted using predictive dynamics is presented in Chapter 9.

The solid black curves in the figure represent the predicted determinants, the dashed curves are the experimental mean value of the determinants, and the curves with shorter dashes show the 95% confidence interval (C.I.) of the statistical means.

The six simulated determinants lie close to the mean of the experimental data; major parts of the motion are in the confidence region and have trends as those of the statistical mean. The peak values of ankle motion, pelvic rotation, and pelvic lateral displacement occur earlier than the experimental data. This is because the six determinants represent a complex coupled motion, and the optimal solution is only a compromise motion. Furthermore, ankle motion is the major passive movement due to GRF. It has shown a larger plantar flexion at 60% of gait cycle. These differences are expected due to approximations in the mechanical model of the human body.

7.11.1.2 Dynamics

Since we are using a rigid skeleton model, the energy absorption of muscle tendons and ligaments and joint tissue are ignored at heel strike and toe-off. This results in some jerks in GRF and joint torques. Therefore, the Butterworth low-pass filter with a cutting frequency of 8 Hz is used to obtain plots for the GRF and joint torques. Figure 7.18 depicts the torque profiles of the hip, the knee and the ankle for a stride. Here, HS denotes heel strike, FF foot flat, HO heel off, and TO toe-off. The figure also shows torques data (digitized) obtained from the literature (Simpson and Jiang, 1999; Stansfield et al., 2006).

In Figure 7.18, the hip torque begins to flex the hip at the heel strike and this torque reaches its maximum extension torque at the terminal stance phase. At the knee joint, the reaction force flexes the knee during the early stance, but the knee torque then reverses into an extension torque. Before the swing phase, the knee is flexed for a second time. The ankle starts with a plantar torque just after heel strike and reverses into a dorsiflexion torque continuously during the stance, reaching its peak at terminal stance and then dropping quickly until toe-off.

Figure 7.19 shows the GRFs for the predicted walking motion. The vertical GRF has a familiar double-peak pattern, and the maximum vertical force is developed soon after heel strike and then again during terminal stance (push-off). In the walking direction, the fore-aft GRF, there is a decelerating force early in the stance phase, and an acceleration force at push-off. Meanwhile, the foot also pushes laterally during the entire stance phase.

The above resulting joint torques and GRF have shown general agreement with the experimental results presented in the literature (Simpson and Jiang, 1999; Stansfield et al., 2006). However, there are discrepancies at the beginning and end of the gait cycle. Since we do not impose symmetry conditions on joint angle accelerations, the predicted forces show impact phenomena with discontinuities at the boundaries. The discrepancies may also be due to the approximate models used and approximate simulation of impacts during walking motion. In addition, the GRF are linearly distributed between the feet during the double support phase and this may also result in some inaccuracies.

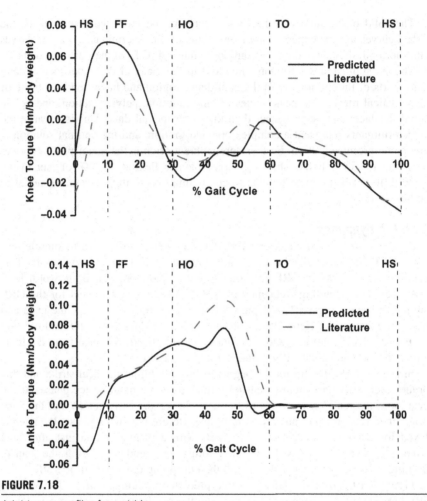

FIGURE 7.18

Joint torque profiles for a stride.

7.12 Cause and effect

The most important aspect of a human simulator is the ability to study tradeoffs, cause and effect. One such aspect is the loading of a human model with various types of loads at various locations. The intent is to see the effect on the human's biomechanics, in this case walking.

Consider the case where humans carry loads on their backs while walking. We shall assume a walking velocity $V = 1.2$ m/s and a step length $L = 0.6$ m. The load is considered as a point load applied on the back in the downward vertical direction. Although this load model is relatively simple, it is used to study cause and effect during the gait cycle. Three cases are simulated with varying load weights of 20 lbs, 40 lbs, and 80 lbs. 3D stick diagrams of the walking motion are compared in Figure 7.20. The walks of Santos[®] using predictive dynamics for no

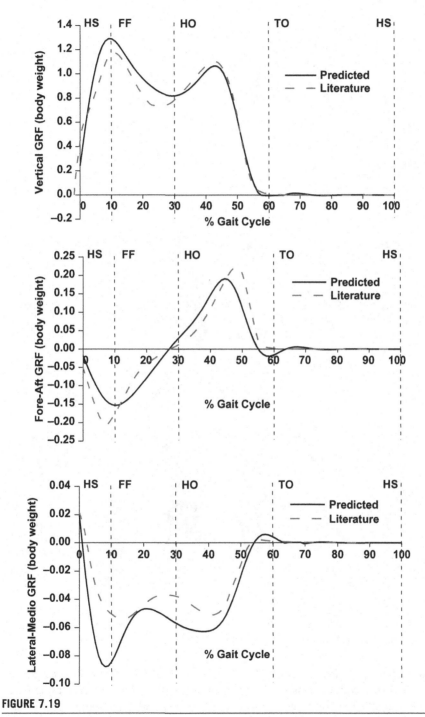

FIGURE 7.19

Ground reaction forces for a stride.

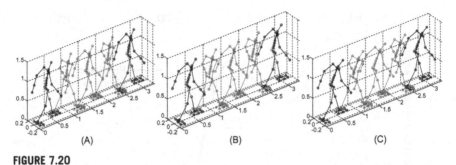

FIGURE 7.20

Cause and effect on walking motion with loads of (A) 20 lb, (B) 40 lb, (C) 80 lb.

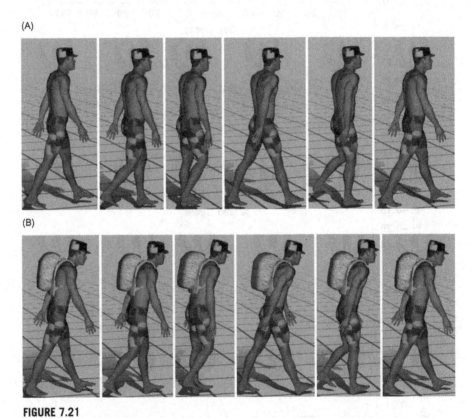

FIGURE 7.21

(A) Snapshots of predictive dynamics results of cyclic walking motion without a backpack; (B) results of predictive dynamics of walking motion with a 60-lb backpack.

external load compared with a load of 60 lbs is shown in Figures 7.21A and 7.21B, respectively. A reasonable spine bending is observed with increasing backpack weight. The joint angle profiles, the GRFs, and joint torque profiles are illustrated in Figures 7.22, 7.23, and 7.24, respectively.

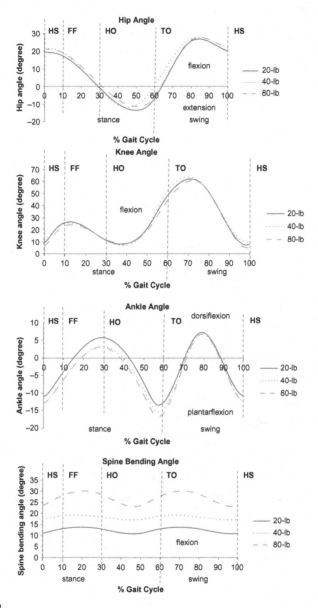

FIGURE 7.22

Joint angle profiles with backpack.

    The hip, knee, and ankle motions have no significant differences with different backpacks. This is because they are all simulated with the same walking velocity and step length. However, the spine bending angles are significantly affected by the backpack weights. Heavier weight results in larger spine bending to lower the center of mass and increase stability.

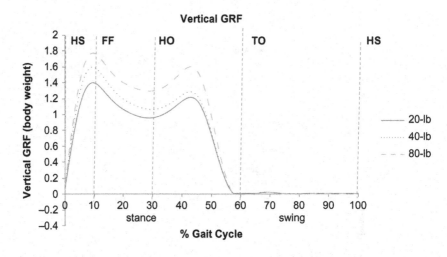

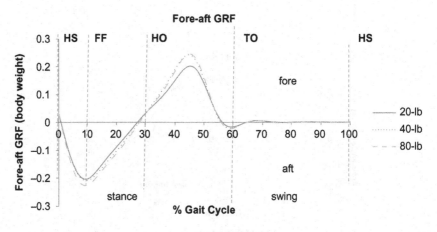

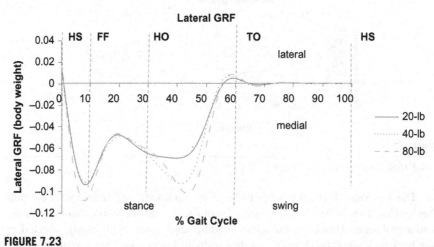

FIGURE 7.23

Ground reaction forces with backpack.

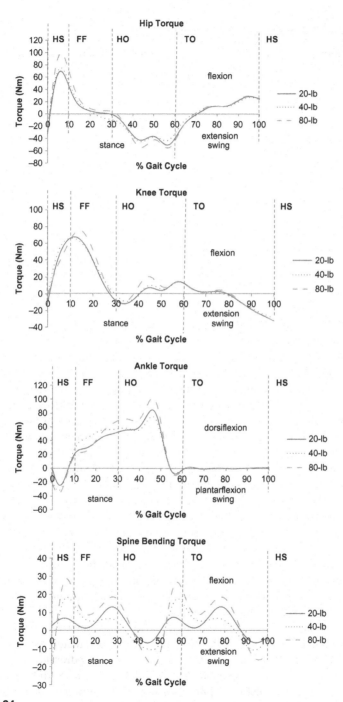

FIGURE 7.24

Joint torque profiles with backpack.

Table 7.3 Major Joint Angle Limits[a,b]

Joints	Joint Angle Limits (degree)	
	Lower Limit	Upper Limit
Ankle (dorsi/plantar)	−20	54.5
Knee (extension/flexion)	7	138
Hip (flexion/extension)	−102	41
Hip (abduct /adduct)	−46	34
Hip (external /internal)	−49	32
Spine (tilt)	−11	11
Spine (bend)	−9.5	21
Spine (rotate)	−13.5	13.5
Shoulder (aft/fore)	−19	111
Shoulder (adduct/abduct)	−23	123.5

[a]Zero joint angles correspond to home configuration as depicted in Figure 7.2
[b]Joint coupling motions are not considered

The GRF show generally larger forces with increasing backpack weight. There is significant backpack effect on the vertical GRF. For the fore-aft GRF, the 80-lb backpack results in greater minimum and maximum force than the 20-lb backpack. The 40-lb backpack has a minimum force similar to the 20-lb backpack and a maximum force similar to the 80-lb backpack. For medial-lateral GRF, the 80-lb backpack has larger peak force compared with the 20-lb and 40-lb backpacks; however, there is no significant difference in other parts of the gait cycle.

It is seen that the torques on the lower extremity are almost the same in the swing phase; however, they are different in the stance phase because of varying backpack weights. In the swing phase, all the weight is shifted to the supporting leg, so the torque of the swinging leg is almost the same with different backpacks. This is quite reasonable. On the other hand, during the stance phase, an 80-lb backpack results in larger peak torque compared with 20-lb and 40-lb backpacks. For the hip and knee torques, there is no significant difference between 20-lb and 40-lb backpacks. The maximum extension knee torque of an 80-lb backpack is similar to those of other backpacks, but the maximum flexion torque occurs later than that for the lighter backpacks. The spine-bending torque shows significantly larger value with increases in backpack weight. Major joint angle limits are presented in Table 7.3 again for convenience.

An interesting result is observed from the foregoing analyses. Different backpack weights are considered with the same walking parameters (step length and velocity). Therefore, similar joint profiles of the lower extremities are obtained with different backpacks. Although the GRF show significant differences, especially for the vertical GRF, the joint torque profiles show no significant differences. This may be explained in the context of human walking strategy; the GRF location is optimized to facilitate an energy-saving walk for different backpacks to increase stability and reduce the joint torques. Thus, the current algorithm predicts that people will choose a strategy to walk more efficiently for carrying backpacks under the given walking parameters.

7.13 Implementations of the predictive dynamics walking formulation

7.13.1 Effect of constrained joints

Constrained joints refer to the range of motion limitations in various joints that can be due to disabilities, injuries, or physical restraints. Examples of physical restraints include clothing of various thickness and stiffness, and rigid objects that can be worn by humans such as knee pads and armor material. The predictive dynamics formulation for walking takes into consideration ranges of motion as limitations on the DOF for each joint. These are imposed as inequality constraints that can be varied at any time.

7.13.2 Sideways and backward walking

Sideways and backward walking are important aspects of walking simulation and are indeed a realistic human behavior.

7.13.3 Effect of changing anthropometry

Because of the implementation of the DH parameters, the dimensions of each body part are readily implemented into the DH table. Therefore, a simulation can include variations in anthropometric measures to see the effect on the motion. Note that any change in anthropometry should be, although not necessarily, accompanied with a change in the set of strength surfaces associated with that simulation. Limits of strength are assigned as presented in Chapter 6.

7.13.4 Effect of changing loads

As demonstrated in the previous section, a change of loading on the human will directly affect the optimization problem. Loads on the arms, legs, torso, or any other body location can be simulated and their effects calculated and visualized. Loading effects are of particular importance for athletes, military personnel, and workers in industrial settings. Analyzing and better understanding the effect of loading on human performance is at the core of predictive dynamics and is why this book was created.

7.13.5 **Walking on uneven terrains**

While only simple forward walking was presented here, the method has been extended to walking on uneven terrain (Hariri, 2011). The major change in the formulation is in the ZMP approach that was originally developed for forward walking on a flat surface. Walking on uneven terrains includes stair climbing and many other implementations.

7.13.6 **Asymmetric walking**

Asymmetric walking can be simulated by formulating the problem for a full stride as demonstrated by Xiang et al. (2009a,b, 2010a,b,c).

7.13.7 **Walking on different terrain types**

Foot contact with the ground, including friction conditions and various ground types such as mud or ice, is important in many fields. The interaction between the foot and soft inelastic material such as gravel is an area of future research.

7.14 **Concluding remarks**

This chapter is dedicated to the detailed treatment of formulating a predictive dynamic walk cycle. The main objective is to demonstrate how gait prediction is formulated and how it can be used to study cause and effect. We summarize the main advantages of using predictive dynamics as a method for understanding motion and the effect of physics in modeling and simulating human motion.

a. Walking with various forms, loads, anthropometries, and ranges of motion can be simulated.
b. Integration of the equations of motion is not carried out. As a result, the prediction of a walk cycle is faster and fundamentally of higher fidelity because of the optimization algorithm.
c. The cause and effect can be obtained by varying the input to the model.
d. A high number of DOFs for the human body (55 DOF were used) can provide an adequate, natural, and realistic motion.
e. Global DOFs and virtual joints can be used to provide for the global transformation motion of the digital human with respect to a fixed reference frame.
f. The objective function used as a characterization of the joint torque square provides a driving function for the motion, to enable a natural human-like realistic behavior.
g. Ground reaction forces using the predictive dynamics formulation can be recovered.

h. The ZMP is readily implemented into the predictive dynamics formulation for walking and provides effective balance criteria for walking.

i. Global forces, reaction forces, joint and torque profile can be calculated from the simulation.

References

Anderson, F.C., Pandy, M.G., 2001. Dynamic optimization of human walking. J. Biomech. Eng.—Trans. ASME 123 (5), 381−390.

Ayyappa, E., 1997. Normal human locomotion, part 1: basic concepts and terminology. J. Prosthet. Orthot. 9 (1), 10−17.

Chevallereau, C., Aoustin, Y., 2001. Optimal reference trajectories for walking and running of a biped robot. Robotica 19, 557−569.

Choi, M.G., Lee, J., Shin, S.Y., 2003. Planning biped locomotion using motion capture data and probabilistic roadmaps. ACM Trans. Graph. 22 (2), 182−203.

Dasgupta, A., Nakamura, Y., 1999. Making feasible walking motion of humanoid robots from human motion capture data. IEEE International Conference on Robotics and Automation, Tokyo, Japan, pp. 1044−1049.

Fregly, B.J., Reinbolt, J.A., Rooney, K.L., Mitchell, K.H., Chmielewski, T.L., 2007. Design of patient-specific gait modifications for knee osteoarthritis rehabilitation. IEEE T. Bio.-Med. Eng. 54 (9), 1687−1695.

Gill, P.E., Murray, W., Saunders, M.A., 2002. SNOPT: an SQP algorithm for large-scale constrained optimization. Siam. J. Optimiz. 12 (4), 979−1006.

Goswami, A., 1999. Postural stability of biped robots and the foot-rotation indicator (FRI) point. Int. J. Robot. Res. 18 (6), 523−533.

Huang, Q., Yokoi, K., Kajita, S., Kaneko, K., Arai, H., Koyachi, N., et al., 2001. Planning walking patterns for a biped robot. IEEE T. Robot. Autom. 17 (3), 280−289.

Kajita, S., Kanehiro, F., Kaneko, K., Fujiwara, K., Harada, K., Yokoi, K., et al., 2003. Biped walking pattern generation by using preview control of zero-moment point. IEEE International Conference on Robotics and Automation Taipei, Taiwan, pp. 1620−1626.

Kim, J., Abdel-Malek, K., Yang, J., Nebel, K., 2005. Optimization-based dynamic motion simulation and energy consumption prediction for a digital human. J. Passenger Car: Electron. Electronical Syst. 114 (7), 797−806.

Lo, J., Huang, G., Metaxas, D., 2002. Human motion planning based on recursive dynamics and optimal control techniques. Multibody Syst. Dyn. 8 (4), 433−458.

Mu, X., Wu, Q., 2003. Synthesis of a complete sagittal gait cycle for a five-link biped robot. Robotica 21, 581−587.

Park, J., Kim, K., 1998. Biped Robot Walking Using Gravity-Compensated Inverted Pendulum Mode and Computed Torque Control. IEEE International Conference on Robotics and Automation, pp. 3528−3533.

Pettre, J., Laumond, J.P., 2006. A motion capture-based control-space approach for walking mannequins. Comput. Animation Virtual Worlds 17 (2), 109−126.

Ren, L., Jones, R.K., Howard, D., 2007. Predictive modelling of human walking over a complete gait cycle. J. Biomech. 40 (7), 1567−1574.

Saidouni, T., Bessonnet, G., 2003. Generating globally optimized sagittal gait cycles of a biped robot. Robotica 21, 199–210.

Sardain, P., Bessonnet, G., 2004. Forces acting on a biped robot. Center of pressure-zero moment point. IEEE T. Syst. Man Cy. A 34 (5), 630–637.

Saunders, J.B.D.M., Inman, V.T., Eberhart, H.D., 1953. The major determinants in normal and pathological gait. J. Bone Joint Surg. Am. 35-A (3), 543–558.

Simpson, K.J., Jiang, P., 1999. Foot landing position during gait influences ground reaction forces. Clin. Biomech. 14 (6), 396–402.

Stansfield, B.W., Hillman, S.J., Hazlewood, M.E., Robb, J.E., 2006. Regression analysis of gait parameters with speed in normal children walking at self-selected speeds. Gait Posture 23 (3), 288–294.

Vukobratović, M., Borovac, B., 2004. Zero-moment point—35 years of its life. Inter. J. HR 1 (1), 157–173.

Xiang, Y., 2008. Optimization-based dynamic human walking prediction (PhD. Dissertation), The University of Iowa, 141 pages.

Xiang, Y., Rahmatalla, S., Bhatt, R., Kim, J., Chung, H.J., Mathai, A., et al., 2008. Optimization-based dynamic human lifting prediction. Paper No. 2008-01-1930. Proceedings of the SAE Digital Human Modeling Conference, Pittsburgh, PA.

Xiang, Y., Arora, J.S., Rahmatalla, S., Abdel-Malek, K., 2009a. Optimization-based dynamic human walking prediction: one step formulation. Inter. J. Numer. Methods Eng. 79 (6), 667–695.

Xiang, Y., Arora, J.S., Rahmatalla, S., Bhatt, R., Marler, T., Abdel-Malek, K., 2009b. Human lifting simulation using multi-objective optimization approach. Multibody Syst. Dyn. 23 (4), 431–451.

Xiang, Y., Arora, J.S., Abdel-Malek, K., 2010a. Optimization-based prediction of asymmetric human gait. J. Biomech. 44 (4), 683–693.

Xiang, Y., Arora, J.S., Abdel-Malek, K., 2010b. Physics-based modeling and simulation of human walking: a review of optimization-based and other approaches. Struct. Multi. Optimiz. 42 (1), 1–23.

Xiang, Y., Chung, H.J., Kim, J.H., Bhatt, R., Rahmatalla, S., Yang, J., et al., 2010c. Predictive dynamics: an optimization-based novel approach for human motion simulation. Struct. Multi. Optimiz. 41 (3), 465–479.

Yamaguchi, J., Soga, E., Inoue, S., Takanishi, A., 1999. Development of a bipedal humanoid robot control method of whole-body cooperative dynamic biped walking. IEEE International Conference on Robotics and Automation, pp. 368–374.

Predictive Dynamics: Lifting

With contributions by Dr. Yujiang Xiang

First they ignore you, then they laugh at you, then they fight you,
then you win.
Mahatma Gandhi (1869–1948)

8.1 Human skeletal model

Lifting is a common task that is performed by humans every day. It is described as holding an object with two hands and moving the object from a lower position to a higher position. There have been several studies involving lifting, as it is a common task that often leads to injury (Arisumi et al., 2007; Dysart and Woldstad, 1996; Huang et al., 2005; Xiang et al., 2009b; Zhang et al., 2000).

For this predictive dynamics task, we shall use the same kinematic skeleton and corresponding DH representation as that shown in Section 7.4.1, Figure 7.2 (shown again in Figure 8.1), with 55 DOFs, and the link lengths and mass properties shown in Table 7.1. The majority of the chapter is adapted from Xiang et al. (2009b, 2012).

8.2 Equations of motion and sensitivities

We shall use the recursive formulation introduced in Chapter 4 to formulate the predictive dynamics task of lifting. Forward kinematics propagates the motion from the base point to the end-effectors. In contrast, the backward dynamics transfer the forces from the end-effectors to the base point.

8.2.1 Forward recursive kinematics

Position, velocity, and acceleration characterized by the 4×4 matrices $\mathbf{A}_j$, $\mathbf{B}_j$, and $\mathbf{C}_j$, respectively, are recursively defined for the jth joint. Given the link transformation matrix, $\mathbf{T}_j$, and the kinematics state variables for each joint, angular displacement, velocity, and acceleration, q_j, $\dot{q}_j$, and $\ddot{q}_j$, we define for $j = 1$ to n as follows:

$$\mathbf{A}_j = \mathbf{T}_1 \mathbf{T}_2 \mathbf{T}_3 \cdots \mathbf{T}_j = \mathbf{A}_{j-1} \mathbf{T}_j \tag{8.1}$$

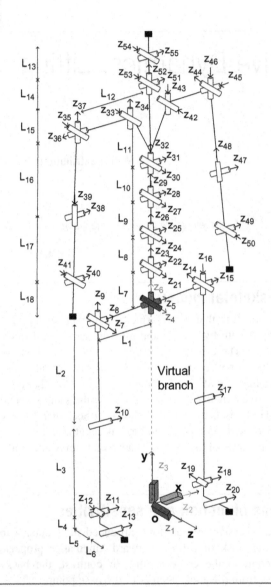

FIGURE 8.1

The 55-DOF digital human model (with global DOFs z_1, z_2, z_3, z_4, z_5, z_6).

$$\mathbf{B}_j = \dot{\mathbf{A}}_j = \mathbf{B}_{j-1}\mathbf{T}_j + \mathbf{A}_{j-1}\frac{\partial \mathbf{T}_j}{\partial q_j}\dot{q}_j \qquad (8.2)$$

$$\mathbf{C}_j = \dot{\mathbf{B}}_j = \ddot{\mathbf{A}}_j = \mathbf{C}_{j-1}\mathbf{T}_j + 2\mathbf{B}_{j-1}\frac{\partial \mathbf{T}_j}{\partial q_j}\dot{q}_j + \mathbf{A}_{j-1}\frac{\partial^2 \mathbf{T}_j}{\partial q_j^2}\dot{q}_j{}^2 + \mathbf{A}_{j-1}\frac{\partial \mathbf{T}_j}{\partial q_j}\ddot{q}_j \qquad (8.3)$$

where $A_0 = 1$ and $B_0 = C_0 = 0$. Then, the global position, velocity, and acceleration of a point in the Cartesian coordinate system can be calculated using the following formulas:

$$^0\mathbf{r}_j = \mathbf{A}_j\mathbf{r}_j; \quad ^0\dot{\mathbf{r}}_j = \mathbf{B}_j\mathbf{r}_j; \quad ^0\ddot{\mathbf{r}}_j = \mathbf{C}_j\mathbf{r}_j \tag{8.4}$$

where $^0\mathbf{r}_j$ and $\mathbf{r}_j$ are global and local augmented coordinates, respectively. It is noted that the repeated index is not summed in the above equations or the equations that follow.

8.2.2 Backward recursive dynamics

We proceed to develop the backward recursion for the dynamic analysis, which is accomplished by defining a 4×4 transformation matrix $\mathbf{D}_i$ and 4×1 transformation vectors $\mathbf{E}_i$, $\mathbf{F}_i$, and $\mathbf{G}_i$ as follows.

Given the external force $\mathbf{f}_k^T = \begin{bmatrix} ^kf_x & ^kf_y & ^kf_z & 0 \end{bmatrix}$ and the moment $\mathbf{h}_k^T = \begin{bmatrix} ^kh_x & ^kh_y & ^kh_z & 0 \end{bmatrix}$ for the link k, defined in the global coordinate system, the joint actuation torques τ_i for $i = n$ to 1 are computed as:

$$\tau_i = tr\left[\frac{\partial \mathbf{A}_i}{\partial q_i}\mathbf{D}_i\right] - \mathbf{g}^T\frac{\partial \mathbf{A}_i}{\partial q_i}\mathbf{E}_i - \mathbf{f}_k^T\frac{\partial \mathbf{A}_i}{\partial q_i}\mathbf{F}_i - \mathbf{G}_i^T\mathbf{A}_{i-1}\mathbf{z}_0 \tag{8.5}$$

$$\mathbf{D}_i = \mathbf{I}_i\mathbf{C}_i^T + \mathbf{T}_{i+1}\mathbf{D}_{i+1} \tag{8.6}$$

$$\mathbf{E}_i = m_i\,{}^i\mathbf{r}_i + \mathbf{T}_{i+1}\mathbf{E}_{i+1} \tag{8.7}$$

$$\mathbf{F}_i = {}^k\mathbf{r}_f\delta_{ik} + \mathbf{T}_{i+1}\mathbf{F}_{i+1} \tag{8.8}$$

$$\mathbf{G}_i = \mathbf{h}_k\delta_{ik} + \mathbf{G}_{i+1} \tag{8.9}$$

where $\mathbf{D}_{n+1} = 0$ and $\mathbf{E}_{n+1} = \mathbf{F}_{n+1} = \mathbf{G}_{n+1} = 0$; $\mathbf{I}_i$ is the inertia matrix for link i; m_i is the mass of link i; $\mathbf{g}$ is the gravity vector; $^i\mathbf{r}_i$ is the location of center of mass of link i in the local frame i; $^k\mathbf{r}_f$ is the position of the external force in the local frame k; $\mathbf{z}_0 = \begin{bmatrix} 0 & 0 & 1 & 0 \end{bmatrix}^T$ for a revolute joint; and δ_{ik} is Kronecker delta.

Where the first term in the equations of motion is the inertia and Coriolis torque, the second term is the torque due to gravity load, the third term is the torque due to external force, and the fourth term represents the torque due to the external moment.

8.2.3 Sensitivity analysis

This is a highly nonlinear programming problem, and optimization lends itself well to calculating reasonable solutions. Accurate sensitivity is a key factor for efficiently achieving an optimal solution for a gradient-based optimization algorithm, such as the sequential quadratic programming (SQP) method.

The derivatives $\dfrac{\partial \tau_i}{\partial q_k}, \dfrac{\partial \tau_i}{\partial \dot{q}_k}, \dfrac{\partial \tau_i}{\partial \ddot{q}_k}$ ($i = 1$ to n; $k = 1$ to n) are evaluated for the articulated spatial human mechanical system in a recursive manner using the foregoing recursive Lagrangian dynamics formulation as follows:

$$\frac{\partial \tau_i}{\partial q_k} = \begin{cases} tr\left(\dfrac{\partial^2 \mathbf{A}_i}{\partial q_i \partial q_k}\mathbf{D}_i + \dfrac{\partial \mathbf{A}_i}{\partial q_i}\dfrac{\partial \mathbf{D}_i}{\partial q_k}\right) - \mathbf{g}^{\mathrm{T}}\dfrac{\partial^2 \mathbf{A}_i}{\partial q_i \partial q_k}\mathbf{E}_i - \mathbf{f}^{\mathrm{T}}\dfrac{\partial^2 \mathbf{A}_i}{\partial q_i \partial q_k}\mathbf{F}_i - \mathbf{G}_i^{\mathrm{T}}\dfrac{\partial \mathbf{A}_{i-1}}{\partial q_k}\mathbf{z}_0 & (k \le i) \\[3mm] tr\left(\dfrac{\partial \mathbf{A}_i}{\partial q_i}\dfrac{\partial \mathbf{D}_i}{\partial q_k}\right) - \mathbf{g}^{\mathrm{T}}\dfrac{\partial \mathbf{A}_i}{\partial q_i}\dfrac{\partial \mathbf{E}_i}{\partial q_k} - \mathbf{f}^{\mathrm{T}}\dfrac{\partial \mathbf{A}_i}{\partial q_i}\dfrac{\partial \mathbf{F}_i}{\partial q_k} & (k > i) \end{cases}$$

$$(8.10)$$

$$\frac{\partial \tau_i}{\partial \dot{q}_k} = tr\left(\frac{\partial \mathbf{A}_i}{\partial q_i}\frac{\partial \mathbf{D}_i}{\partial \dot{q}_k}\right) \tag{8.11}$$

$$\frac{\partial \tau_i}{\partial \ddot{q}_k} = tr\left(\frac{\partial \mathbf{A}_i}{\partial q_i}\frac{\partial \mathbf{D}_i}{\partial \ddot{q}_k}\right) \tag{8.12}$$

Note that the computational cost of the recursive formulation is of the order $O(n)$, where n is the number of DOFs. Forward kinematics transfers the motion from the origin toward the end-effector along the branch. In contrast, the backward dynamics propagates forces from the end-effector to the origin. More details about the derivation of sensitivity equations are provided by Xiang et al. (2009a,b).

8.3 Dynamic stability and ground reaction forces (GRF)

We shall use the ZMP method to address the dynamic stability condition for the lifting motion. We shall also include the GRF to be calculated. This can be accomplished by forcing the feet to stay in the form of a polygon as shown in Figure 8.2. The two feet are fixed on the ground with the distance d and orientation angle θ during the lifting motion. The concept of ZMP has been extensively used as a bipedal dynamic stability criterion. It is defined as the point on the ground at which the resultant tangential moments are zero.

We shall also use an active-passive algorithm to calculate ZMP and GRF to obtain the real joint torques for the multi-body human system. Details of the algorithm are presented by Xiang et al. (2009a,b) and Hariri (2012) and outlined as follows:

1. Use inverse dynamics to calculate the global resultant active forces, $\mathbf{M}^o$, $\mathbf{F}^o$, at the origin in the inertial reference frame (o-xyz in Figure 8.2). Note that the state variables q, $\dot{q}$, $\ddot{q}$ (design variables) are specified for each DOF.
2. Calculate the ZMP position from its definition using the global resultant active forces as follows:

$$y_{zmp} = 0; \quad x_{zmp} = \frac{M_z^o}{F_y^o}; \quad z_{zmp} = \frac{-M_x^o}{F_y^o} \tag{8.13}$$

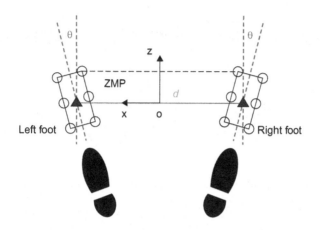

FIGURE 8.2

Feet locations and support region.

where $M^o = \begin{bmatrix} M^o_x & M^o_y & M^o_z \end{bmatrix}^T$ and $F^o = \begin{bmatrix} F^o_x & F^o_y & F^o_z \end{bmatrix}^T$. Note that the two feet are assumed to be on the level ground.

3. Calculate the resultant active forces at ZMP (M^{zmp}, F^{zmp}) using the equilibrium condition as follows:

$$M^{zmp} = M^o + F^o \times {}^o r_{zmp}$$

$$F^{zmp} = F^o \tag{8.14}$$

where ${}^o r_{zmp}$ is the ZMP position in the global coordinate system obtained from Equation (8.13).

4. Calculate the value and location of GRF from the equilibrium between the resultant active forces and passive forces at the ZMP:

$$M^{GRF} + M^{zmp} = 0$$

$$F^{GRF} + F^{zmp} = 0 \tag{8.15}$$

$${}^o r_{GRF} - {}^o r_{zmp} = 0$$

Next, the resultant GRF is partitioned into two feet by using a linear distance relationship between the GRF location and feet centers as shown in Figure 8.3.

5. Finally, all active forces (gravity, inertia, and external forces) and passive forces (GRF) are applied to the multi-body human system using Equation (8.5) to obtain the real joint torques that are used in the torque limit constraints and to calculate the dynamic effort objective function.

8.4 Formulation

8.4.1 Lifting task

In this chapter, the lifting task is defined as moving a box from an initial location to a final location. Figure 8.4 depicts the input parameters for the proposed

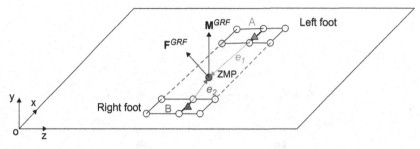

FIGURE 8.3

Partition of the GRF into two feet: e_1 and e_2 are distances between ZMP and feet ball centers.

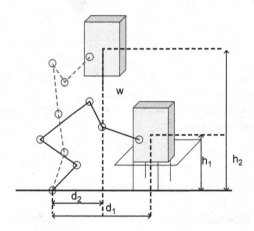

FIGURE 8.4

Definition of input parameters for lifting task.

formulation. In this regard, h_1 is the initial height of the box measured from the ground, d_1 is the initial distance measured from the foot location to the center of the box, h_2 is the final height measured from the ground, d_2 is the final distance, and w is the weight of the box. The initial and final postures and dynamic lifting trajectory are determined by solving a nonlinear optimization problem. The mechanical system is at rest at the initial and final time points.

8.5 Predictive dynamics optimization formulation

The lifting motion is predicted by solving a multi-objective optimization problem. In the proposed formulation, the box initial and final locations, feet locations and orientations, and box dimension and weight are given. The total time T for the

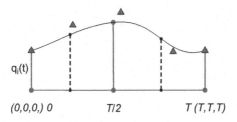

FIGURE 8.5

B-spline discretization of a joint profile.

lifting motion is specified. The initial and final postures are optimized along with the lifting motion instead of specifying them from the experiments.

8.5.1 Design variables and time discretization

Because we must calculate the joint profile across time, we will discretize the time domain using a B-spline. The time domain is discretized by using cubic B-spline functions. Thus a joint profile $q(t)$ is parameterized as follows:

$$q_j(\mathbf{t}, \mathbf{P}) = \sum_{i=1}^{m} N_i(\mathbf{t})p_i \quad 0 \le t \le T \tag{8.16}$$

where $N_i(\mathbf{t})$ represents the basis functions; $\mathbf{t} = \{t_0, ..., t_s\}$ is the knot vector; and $\mathbf{P}_j = \{p_1, ..., p_m\}$ is the control points vector. With this representation, the control points become the optimization design variables. In this study, the knot vector is specified and fixed in the optimization process.

We formulate the lifting task as a general nonlinear programming (NLP) problem. To find the optimal control points $\mathbf{P}$ for the lifting motion, a human performance measure, $F(\mathbf{P})$, is minimized subject to physical constraints as follows:

$$
\begin{aligned}
&\text{Find:} \quad \mathbf{P} \\
&\text{To:} \quad \min F(\mathbf{P}) \\
&\text{Sub.} \quad h_i = 0, \quad i = 1, ..., m \\
&\qquad\quad\; g_j \le 0, \quad j = 1, ..., k
\end{aligned}
\tag{8.17}
$$

where h_i are the equality constraints and g_j are the inequality constraints.

We represent the joint angle profile for each DOF by five control points. As a result, there are 275 design variables (55 DOF × 5 control points). In addition, the total time duration is discretized into four evenly distributed segments, and five time grid points are used for the entire motion as shown in Figure 8.5, where the horizontal scale shows the knot vector. Multiplicity at the ends is used in the knot vector. For B-splines, the multiplicity property guarantees that the initial and final joint angle values of a DOF are exactly those corresponding to the initial and final control point values. The time-dependent constraints are imposed not only at the knot points but also between the adjacent knots, so that a smooth motion can be generated.

Since q, $\dot{q}$, and $\ddot{q}$ are functions of $\mathbf{t}$ and $\mathbf{P}$, torque $\tau = \tau(\mathbf{t}, \mathbf{P})$ is an explicit function of the knot vector and control points. Thus, the derivatives of a torque with respect to the control points can be computed using the chain rule as

$$\frac{\partial \tau}{\partial P_i} = \frac{\partial \tau}{\partial q}\frac{\partial q}{\partial P_i} + \frac{\partial \tau}{\partial \dot{q}}\frac{\partial \dot{q}}{\partial P_i} + \frac{\partial \tau}{\partial \ddot{q}}\frac{\partial \ddot{q}}{\partial P_i} \tag{8.18}$$

Equations (8.10−8.12) and (8.18) are used to calculate accurate gradients for the optimization formulation to improve the computational efficiency.

8.5.2 Objective functions

For the dynamic lifting motion prediction problem, two performance measures are investigated. The first one is to minimize the dynamic effort, which is defined as the time integral of the squares of all joint torques; the second one is to maximize the stability, which can be transformed to minimize the time integral of the distance squares between ZMP and the foot support boundaries. The weighted sum of the two objective functions is used as the performance measure as follows:

$$F(\mathbf{P}) = w_1 N\left(\int_{t=0}^{T} \boldsymbol{\tau}(\mathbf{P}, t)^{\mathrm{T}}\boldsymbol{\tau}(\mathbf{P}, t)\, dt\right) + w_2 N\left(\int_{t=0}^{T}\left(\sum_{i=1}^{nb} s(t)_i^2\right) dt\right) \tag{8.19}$$

where s_i is the distance between ZMP and the ith foot support boundary at time-t; nb is the number of foot support boundaries; w_1 and w_2 are weighting coefficients for the two objective functions ranging from 0 to 1, respectively; and $w_1 + w_2 = 1$. The symbol $N(\bullet)$ is a normalization operator. A general function-transformation method (Marler and Arora, 2005) is used to determine the normalization operator in Equation (8.19).

8.5.3 Constraints

We shall consider two types of constraints for the predictive dynamics task of lifting. One type is the time-dependent constraints, which include joint limits, torque limits, ground penetration, dynamic balance, foot locations, vision, hand orientation, and collision avoidance. These constraints are imposed throughout the time interval. The second type is time-independent constraints, which comprise the initial and final box locations and the initial and final static conditions; these constraints are considered only at the starting and ending time points for the lifting motion.

For the lifting task, joint angle limits, torque limits, ground penetration, foot locations, and dynamic balance constraints are detailed by Xiang et al. (2009a,b), and a symmetric walking motion is simulated using a one-step formulation. The vision, hand orientation, collision avoidance, initial and final box locations, and

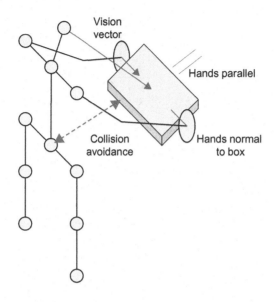

FIGURE 8.6

Hand-orientation, vision, and collision-avoidance constraints.

initial and final static conditions are new constraints for the lifting problem. These new constraints are depicted in Figure 8.6 and described as follows.

8.5.3.1 Vision

We shall enforce a vision constraint, which is to align the eye vector, $\mathbf{r}_{eye}$, with the vision vector, $\mathbf{r}_{vision}$. The vision vector is aligned to the box center. The vision vector is defined as a vector from the midpoint of the two eyes to the box center. The eye vector is defined as a vector located at the midpoint of the eyes and normal to the forehead as shown in Figure 8.7.

$$\alpha \le \theta = \arccos\left(\frac{\mathbf{r}_{eye} \cdot \mathbf{r}_{vision}}{\|\mathbf{r}_{eye}\| \|\mathbf{r}_{vision}\|}\right) \le \beta \qquad (8.20)$$

where α and β define the vision cone.

8.5.3.2 Hand distance and orientation

Hand distance during the lifting task is a constant defined by the box dimension. Furthermore, the hand orientation is represented by two vectors: the hand normal vector, $\mathbf{r}_n$, and the hand tangential vector, $\mathbf{r}_t$, as depicted in Figure 8.8.

$$\mathbf{r}_n \times \mathbf{r}_{box} = 0$$
$$\mathbf{r}_t \times \mathbf{r}_{handle} = 0 \qquad (8.21)$$

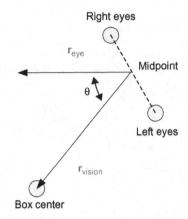

FIGURE 8.7

Vision constraint.

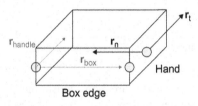

FIGURE 8.8

Hand-orientation constraint.

Note that the orientation constraints are imposed in such a way that the hand normal vector is parallel to the box edge vector, $\mathbf{r}_{box}$, and the hand tangential vector is parallel to the box handle vector, $\mathbf{r}_{handle}$.

8.5.3.3 Collision avoidance

There are many methods for detecting and avoiding collision during the calculation process. A fast and efficient method is to fill objects in the scene with spheres, including the human body. The size of the spheres will determine the accuracy of the collision detection/avoidance algorithm.

Specifically, the ankle, knee, hip, shank, thigh, chest, and neck are filled with spheres of various radii to represent the body dimensions. The distances between the box edge and all sphere centers are calculated to impose collision avoidance as shown in Figure 8.9. Collision avoidance is considered in the current formulation to prevent penetration of the box in the body.

$$l = \frac{\|\mathbf{r}_{body} \times \mathbf{r}_{edge}\|}{\|\mathbf{r}_{edge}\|} \geq r \qquad (8.22)$$

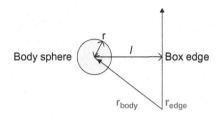

FIGURE 8.9

Collision avoidance constraint between the box and the body.

where r is the radius of the sphere to represent a body segment, and l is the distance between the sphere center and the box surface.

8.5.3.4 Initial and final hand positions

Hand-contacting positions are specified at the initial and final times for the lifting motion. It is noted that the initial and final postures are determined by the optimization process instead of specifying them from experiments.

$$
\begin{aligned}
\mathbf{x}(0) &= \tilde{\mathbf{x}}(0), \\
\mathbf{x}(T) &= \tilde{\mathbf{x}}(T),
\end{aligned}
\tag{8.23}
$$

where $\mathbf{x}$ is the calculated hand position, and $\tilde{\mathbf{x}}$ is the specified hand-contacting position.

8.5.3.5 Initial and final static conditions

The entire body is at rest at the initial and final time points. These conditions are implemented as equality constraints as follows:

$$
\begin{aligned}
\dot{q}_i(0) &= 0, \\
\dot{q}_i(T) &= 0, \quad i = 1, ..., n
\end{aligned}
\tag{8.24}
$$

where n is the number of DOF.

8.6 Computational procedure for multi-objective optimization

A sequential quadratic programming (SQP) algorithm is typically used to solve the nonlinear optimization problem of lifting motion prediction. To use the algorithm, cost and constraint functions and their gradients must be calculated. The recursive kinematics and dynamics provide accurate gradients to improve the computational efficiency of the optimization algorithm (Lee et al., 2005; Xiang et al., 2009a,b). The adaptive lifting strategies are predicted for box lifting by

Table 8.1 Task Parameters for the Box Lifting

Parameters	
Box weight (kg)	9.0
Box width (m)	0.525
Box height (m)	0.365
Box depth (m)	0.370
d_1 (m)	0.490
h_1 (m)	0.365
d_2 (m)	0.460
h_2 (m)	1.370
T (s)	1.2

solving the NLP problem. The data related to the box-lifting task are obtained from an experiment as depicted in Table 8.1.

8.6.1 Lifting determinants and error quantification

As indicated in the formulation for predictive dynamics, motion determinants must be delineated. Based on the literature (Authier et al., 1996) and experimental observations, six joint angle profiles are chosen as the determinants to define the lifting motion. These determinants correspond to the whole-body motion parameters that have major roles in the task: hip flexion/extension, knee flexion/extension, ankle plantar/dorsiflexion, trunk flexion/extension, shoulder flexion/extension, and elbow flexion/extension.

Errors of these determinants between simulation and experiments are quantitatively studied. In general, the error for the dynamic system at time point t is given as:

$$e_i(\mathbf{q}, t) = q_i(t) - \tilde{q}_i(t) \tag{8.25}$$

where $q_i(t)$ is the simulated ith joint angle profile and $\tilde{q}_i(t)$ is the measured ith joint angle profile at time t. Next, the error throughout the time interval T is given in scalar form using an L_2 norm, which is defined as the following integral:

$$E_i(\mathbf{q}) = \int_0^T [e_i(\mathbf{q}, t)]^2 dt \tag{8.26}$$

For box-lifting validation, the total error summation of the six determinants is evaluated as:

$$E_{sum}(\mathbf{q}) = E_{ankle}(\mathbf{q}) + E_{knee}(\mathbf{q}) + E_{hip}(\mathbf{q}) + E_{trunk}(\mathbf{q}) + E_{shoulder}(\mathbf{q}) + E_{elbow}(\mathbf{q}) \tag{8.27}$$

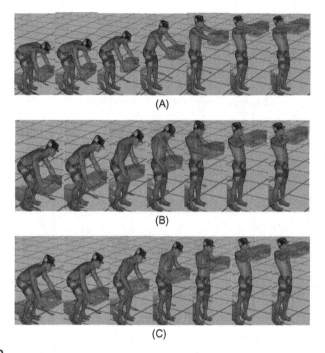

FIGURE 8.10

Snapshots of simulated box-lifting motion: (A) stability criterion, (B) dynamic effort criterion, (C) a weighted sum of $w_1 = 0.1$ and $w_2 = 0.9$.

The total error in Equation (8.27) is used to quantify accuracy of the predicted lifting motion and to determine weighting coefficients corresponding to each objective function for the multi-objective optimization problem.

8.7 Predictive dynamics simulation

For the results below, a weighted sum approach to multi-objective optimization can be used for stability and effort. The box lifting is simulated by solving the MOO problem, and the simulation results are validated with the experimental data in this section. The effects of different objective functions are also illustrated.

Results of the predictive dynamics simulation for the lifting task using stability, dynamic effort, and the weighted sum performance measures are depicted in Figure 8.10. The ZMP trajectories for the box-lifting simulations using different performance measures are illustrated in Figure 8.11. In addition, GRF and joint torque profiles obtained from the motion prediction using the predictive dynamics approach are shown in Figures 8.12 and 8.13, respectively.

A detailed reference on using predictive dynamics in simulating lifting was reported by Xiang et al. (2009b). As observed in Figure 8.10, the stability

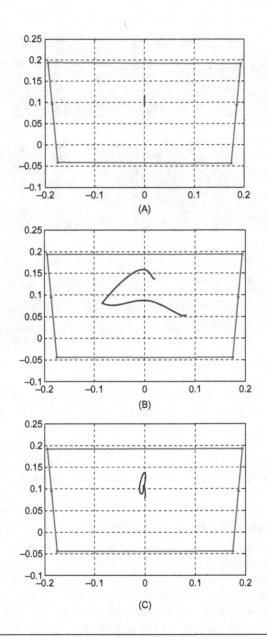

FIGURE 8.11

ZMP and feet support polygon for box lifting with various performance measures:
(A) stability criterion, (B) dynamic effort criterion, (C) weighted sum $w_1 = 0.1$ and $w_2 = 0.9$.

criterion predicts a back-lift box-lifting motion in which a large spine flexion
occurs without obvious knee-bending as shown in Figure 8.10(A). The predicted
lifting motion gives maximum stability, and the ZMP trajectory focuses around
the center of the support polygon in Figure 8.11(A). In contrast, the dynamic

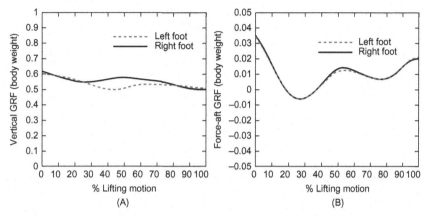

FIGURE 8.12

Ground reaction forces for dynamic box-lifting using the weighted sum performance measure: (A) vertical GRF, (B) fore-aft GRF.

effort performance measure predicts a more natural back-lift lifting motion as shown in Figure 8.10(B). Knee, spine, and shoulder flexions occur simultaneously, and the ZMP trajectory moves in a relatively larger area in the support polygon in Figure 8.11(B). Finally, the weighted sum performance measure produces a lifting motion that is quite similar to the one predicted using the dynamic effort as the performance measure as seen in Figure 8.10(C). However, the predicted ZMP trajectory varies in a small area in the support polygon in Figure 8.11(C). It is evident that the MOO approach shows the combination effects of the two objective functions and also predicts a more natural lifting motion.

The GRF and joint torque profiles for the box-lifting motion are also obtained by solving the optimization problem. The GRFs on both feet are similar in both trends and magnitudes, as shown in Figure 8.12 (GRF is shown as a fraction of the body weight). In addition, the vertical GRF on each foot takes about half of the body and box weights. The fore-aft GRF has small values for a symmetric box-lifting motion. The predicted joint torques for the right knee, hip, shoulder, and elbow are also similar to the values on the left counterparts as depicted in Figure 8.13. This is quite reasonable because the initial and final box locations are symmetric in the sagittal plane so that the predicted lifting motion is also symmetric. However, the torque values on the left and right sides of the skeletal model are not exactly the same. This is expected because of the uneven locations of ZMP where GRF are applied as shown in Figure 8.11(C).

8.8 Validation

In order to ensure the accuracy of the proposed method, motion capture is used to validate the results experimentally. The lifting motion experimental data were collected from five healthy male subjects. The mean height of the subject population was 5′7″ with a mean weight of 143 lbs. The average age of the participants was

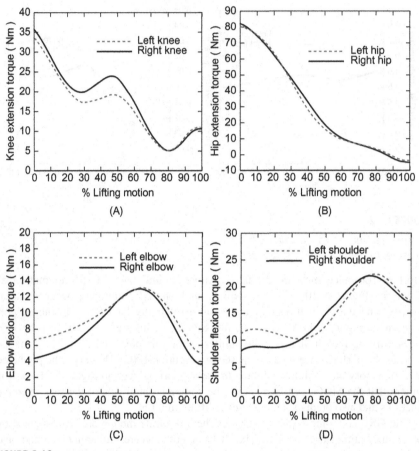

FIGURE 8.13

Joint torque profiles for dynamic box-lifting using weight sum performance measure: (A) knee extension torque, (B) hip extension torque, (C) elbow flexion torque, (D) shoulder flexion torque.

34 years. The subjects had no history of musculoskeletal problems and were reasonably fit. Their participation was voluntary, and a written informed consent, as approved by the University of Iowa Institutional Review Board, was obtained prior to testing. During the lifting trials, each subject lifted at a self-selected speed. The experimental data for each subject were also normalized by dividing the cycle time, which is defined as the time duration to lift the object from the initial position to the final position, to directly evaluate the determinants at a percentage of a lifting cycle. For each subject, the timescale was normalized such that the initial posture occurred at time $t = 0$ and the final posture occurred at time $t = 1$. The lifting motion is validated with experimental determinants as shown in Figure 8.14.

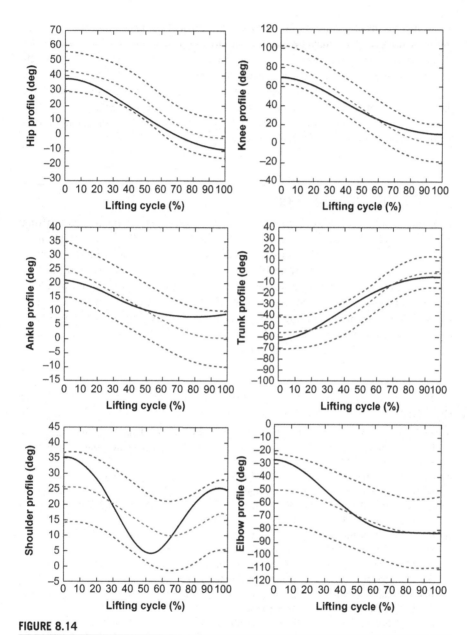

FIGURE 8.14

Comparison of lifting determinants between simulation and experiments.

In general, the predicted lifting motion has shown a good correlation with the experimental data. First, they stay inside the normal region specified by the interval of confidence (outer dashed curves). Additionally, they show similar trends to those of the mean (the middle dashed curve) for the normal subjects.

Nevertheless, the dynamic model shows some difficulties predicting the shoulder angle accurately. This may be due in part to the absence of the necessary constraints on the complex shoulder motion in the model and in part to the difficulties associated with computing accurate shoulder joint motion from the experimental data. The latter could be contributed to the complex structure of the shoulder joint and the high degree of coupling with the adjacent joints in the human body during the lifting motion. It is noted that the validation for the kinetics data is ongoing and will be reported later.

8.9 Concluding remarks

This chapter has presented an implementation of predictive dynamics as a lifting task. The method is broadly applicable to lifting a load from one location to another. The motion planning was formulated as a large-scale nonlinear programming problem. Joint profiles were discretized using cubic B-splines, and the corresponding control points were treated as unknowns for the optimization problem. Two objective functions were used in the lifting formulation: dynamic effort and stability as a weighted sum approach in a multi-objective optimization problem. Based on the simulation data, the ability of the proposed methodology to select a realistic human lifting strategy was demonstrated.

References

Arisumi, H., Chardonnet, J.R., Kheddar, A., Yokoi, K., 2007. Dynamic lifting motion of humanoid robots. 2007 IEEE International Conference on Robotics and Automation, Roma, Italy, pp. 2661–2667.

Authier, M., Lortie, M., Gagnon, M., 1996. Manual handling techniques: Comparing novices and experts. Int. J. Ind. Ergonom. 17 (5), 419–429.

Dysart, M.J., Woldstad, J.C., 1996. Posture prediction for static sagittal-plane lifting. J. Biomech. 29 (10), 1393–1397.

Hariri, M., 2012. A Study of Optimization-Based Predictive Dynamics method for Digital Human Modeling, PhD Dissertation. The University of Iowa, Iowa City, IA.

Huang, C., Sheth, P.N., Granata, K.P., 2005. Multibody dynamics integrated with muscle models and space-time constraints for optimization of lifting movements. ASME International Design Engineering Technical Conferences. ASME, Long Beach, California, Paper No. DETC2005-85385.

Lee, S.H., Kim, J., Park, F.C., Kim, M., Bobrow, J.E., 2005. Newton-type algorithm for dynamics-based robot movement optimization. IEEE Trans. Rob. 21 (4), 657–667.

Marler, R.T., Arora, J.S., 2005. Function-transformation methods for multi-objective optimization. Eng. Optimiz. 37 (6), 551–570.

Xiang, Y., Arora, J.S, Rahmatalla, S., Abdel-Malek, K., 2009a. Optimization-based dynamic human walking prediction: one step formulation. Int. J. Numer. Methods Eng. 79 (6), 667–695.

Xiang, Y., Arora, J.S., Rahmatalla, S., Bhatt, R., Marler, T., Abdel-Malek, K., 2009b. Human lifting simulation using multi-objective optimization approach. Multibody Syst. Dyn. 23 (4), 431−451.

Xiang, Y., Arora, J., Abdel-Malek, K., 2012. 3D human lifting motion prediction with different performance measures. Int. J. Humanoid Robotics, 9 (2), 1250012.

Zhang, X., Nussbaum, M.A., Chaffin, D.B., 2000. Back lift versus leg lift: an index and visualization of dynamic lifting strategies. J. Biomech. 33 (6), 777−782.

Validation of Predictive Dynamics Tasks

9

With Contributions by Salam Rahmatalla, PhD, The University of Iowa

The biggest place I look for validation is from my mother. That's the little girl in me that will never grow up.
Naomi Watts

9.1 Introduction

This chapter introduces a validation methodology to assess the predicted motion of computer human models and provide feedback to model developers for evaluation and refinement. Due to the varied strategies for (and the relative complexity of) human motions under different dynamic tasks, the proposed validation methodology is designed to be general and effective.

While objective and subjective statistical tools comprise the major components of the proposed validation methodology, accurate collection of human data and efficient handling of the data during post-processing operations are critical components in this process. In order to demonstrate the validation method, we will implement it in the two tasks detailed in Chapters 7 and 8.

Advances in measurement systems and motion capture technologies have played a key role in the development of human modeling and simulation. Human models have become more human-like with more sophisticated skeletons and predictive capabilities. As a result, many questions have been raised about the validity of these predictive models in representing human motions. One major question is how to define the line between acceptable and unacceptable predicted motion; there are many possible answers, depending on the required accuracy and application. The objective of this chapter is to present effective validation methodologies that provide tools to answer such questions.

Some approaches for validating predictive human-model motion have been very helpful in the development and acceptance of the models (Abdoli-Eramaki et al., 2009; Blankevoort and Huiskes, 1996; Chaffin, 2002, 2009; Dubowsky et al., 2008; Karduna et al., 2001; Marras and Sommerich, 1991; Rabuffetti and Baroni, 1999; Robert et al., 2007). However, most of these approaches were designed for specific usage or tasks, and they usually targeted a certain area of the human body or used a low-fidelity model with a limited number of links and joints.

A major challenge for validation of predictive dynamics (PD) tasks is the uncertainty in the predicted motion because people assume different strategies when performing the same task. This is attributed to differences in strength and anthropometry, but also to other physiological and psychological parameters that are beyond the scope of this book. The variances in strategies will produce different joint angle profiles for various people. As a result, it is expected that validation processes that are based on only statistical measures may not be effective in capturing such differences.

The second difficulty is associated with the technology for measuring and capturing motion, often referred to as motion capture. For example, a digital human modeled as having 55 degrees of freedom (DOF), with each DOF having a joint profile that must be recorded over a time period, may generate an amount of data that is difficult to handle. Fortunately, in most cases, the real active motion is conducted by a limited number of DOF that represent a subset of the total DOF. The motion of the remaining DOF could be considered as passive, and may not be critical to the validation process.

Another potential difficulty in the validation of PD tasks is the accuracy of collecting the experimental data and transforming them from the Cartesian space to the joint space. In this regard, state-of-the-art motion capture systems with appropriate calibration and a realistic number of cameras should be used to acquire the subject's motion data using appropriate marker placement protocols. Robust inverse-kinematic software should be used in calculating the joint angles from the positional marker data. There are several inverse kinematics (IK) programs to do that; however, most of them are commercial packages and are limited to animating their avatars based on Cartesian information. They may not be useful for animating the predictive human model under investigation (Santos) because it has a detailed human-like skeleton.

The last, but not least, major difficulty in the validation of motion is human perception of virtual reality, which comes into effect when people from different backgrounds are asked to evaluate the motion of an avatar. Two people observing the same avatar may have different impressions of the realism of the avatar's motion; this happens even when the avatar's motion is derived with accurate motion capture systems from human subjects. The role of human perception of virtual reality will become an important component of the validation methodology when it comes to discussing what is acceptable motion and what is not.

In this chapter, a framework to validate the predicted motion of a whole-body task is introduced. The validation framework is based on benchmark tests to characterize the conditions under which the predicted motions are considered acceptable. Some of the benchmark tests are based on qualitative comparisons and are used to construct a general perception about the normality of the motion. The validation method and process should be thought of as a mechanism for providing feedback as well as substantial insight into the PD task.

Additional benchmark tests are based on quantitative comparisons and provide critical and detailed information about the quality of the model in general and the weaknesses at specific local locations. The tests also impose tighter constraints for the model testing and improvement. The validation of two PD tasks—normal walking and box lifting—will be detailed in this chapter.

9.2 **Motion determinants**

The concept of motion determinants is characterized by a set of motion profiles that are subtended by the human body and that are common to humans conducting the same motion.

In this work we will call the active DOF the determinants of the motion. Note that these determinants are well established in gait analysis, but for all other tasks we have introduced an analogous concept to ascertain the quantities that can be measured and validated.

The motion determinants concept presents a very useful time history of the trend of the major joints; it also generates large amounts of motion information. Luckily, the determinants have distinctive signatures at certain locations in the time history of the motion. For example, in normal walking, the knee flexion, as one of the walking determinants, reaches a maximum value of 62 ± 6 degrees during the walking cycle (Baliunas et al., 2002). Such key frames should be selected carefully and used as signatures for normal motion. Fortunately, critical key frames are well defined in the literature for some tasks, such as normal walking; however, they need to be characterized for other tasks.

9.3 **Motion capture systems**

9.3.1 **Overview**

There are many motion capture systems on the market that can be used to acquire the motion characteristics of a human with a relatively high degree of accuracy. These include inertial, optical, and markerless motion capture systems. Each system has its own advantages and disadvantages.

Inertial systems such as those by Xsens (http://www.xsens.com/) and Animazoo (http://www.animazoo.com/) can acquire accurate local acceleration data at different locations on the human body, but they suffer from motion drifting because they lack a global coordinate system; also, they are sensitive to magnetic fields.

Optical motion capture systems can be categorized as passive (reflective markers), such as those by Vicon (http://www.vicon.com/) and Motion Analysis (http://www.motionanalysis.com/), or active (light emitting diodes [LED]), such as those by Optotrak (http://www.ndigital.com/lifesciences/certus-motioncapturesystem.php). Passive motion capture systems have high initial and running costs, but they are still the most popular in international motion research labs because of their accuracy.

Another advantage is that the markers are passive sensors, meaning that they are only reflective surfaces and can be attached easily to any area on the body of the subject without the need for wires to connect them to a data collection system. In addition, theoretically, only three markers are required to define the three-dimensional (3D) velocity and acceleration of each body segment. Another advantage is that there are many well-accepted professional packages, such as Visual3D by C-Motion (http://www.c-motion.com/products/visual3d/) that can be used to post-process the motion capture data for general and specific applications. Finally, real-time data flow and avatar animation can be done easily with most optical motion capture systems, and there are numerous post-processing options to digitally filter and differentiate the motion data.

A potential problem with passive markers, though, is *occlusion*, where the markers do not appear in enough of the camera shots due to blockage of the line of sight by objects in the scene or by other parts of the subject's body. Most commercial post-processing packages have the capability to deal with occluded markers. They may create virtual markers to help fill in the occluded markers, or they may use redundant markers (more than the minimum required in the standard protocol) to compensate for occluded markers. Methodologies are available to fill in the motion data from occluded markers with information from the redundant markers.

Markerless motion capture systems such as those by organic motion (http://www.organicmotion.com/) may present an alternative new approach to motion capture technology; however, their validity as an effective tool still needs to be shown.

9.3.2 Optical motion capture systems

A 12-camera Vicon System (infrared SVcam cameras with a resolution of 0.3 megapixels per frame and a peak capture rate of 200 Hz) and a 16-camera Motion Analysis system (with four megapixels per frame and a peak capture rate of 500 Hz) are used for data collection at the 3D Bio-Motion Research Lab (3DBMRL) at the Center for Computer-Aided Design at The University of Iowa. The basic idea behind the function of optical systems is that each camera in the system sends infrared light that will be reflected back when it collides with reflective surfaces. The Cartesian position of the center of a reflective marker (normally with a spherical or hemispherical shape) can be identified by the intersection of the infrared lights of three cameras. Therefore, with more cameras, there is a better chance that a marker will be seen by more than three cameras during the experiments.

Figure 9.1 illustrates the procedure of using the motion capture system to acquire human motion data for a person wearing a motion capture suit with a number of markers attached to their body. The placement of the markers on the human body is called a marker protocol. There are several standard marker protocols, such as the plug-in gait for gait analysis, that are used in various labs, and

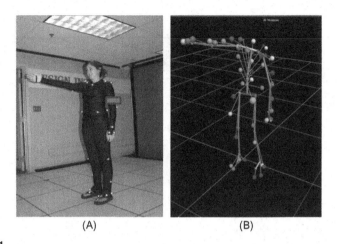

(A)   (B)

FIGURE 9.1

(A) A participant wearing a motion capture suit; (B) Vicon skeletal human model.

new marker protocols can be generated and used for different applications and for avatars with different skeletons. For example, this may be necessary for PD tasks where the motion can be very complicated when a subject is wearing or holding different equipment.

9.3.3 Marker placement protocol

Figure 9.2 depicts the marker protocol that is used for the PD model when preparing the human subjects for the motion capture process (Santos® marker protocol). In this protocol, markers are placed on the subjects to highlight bony landmarks and identify segments between joints based on previously identified guidelines and suggestions (Dubowsky et al., 2008; Karduna et al., 2001). The skeleton of Santos® includes the major joints of the human body with the number of spine joints reduced to four.

Reflective markers are placed on the subject's body to highlight anatomical landmarks. Head markers are placed just superior and lateral to each eyebrow, on each side of the back of the head, and laterally on the level of C1 over the mastoid process; trunk markers are located on the C7, T7, T12, L3, and S1 spinous processes, the jugular notch between the clavicles, each clavicle midway between the manubrium and acromion, the xyphoid process, and the anterior and posterior superior iliac spine. Lower-extremity markers are placed lateral to each greater trochanter, over the medial and lateral condyles of the femur, over the midpoint of the patella, over the medial malleolus of the tibia, over the lateral malleolus of the fibula, just proximal to the 5th metatarsal head, and over the head of the 1st metatarsal. Upper-extremity markers are placed over the acromion process, on the anterior and posterior aspects of the shoulder midway between the lateral edge of

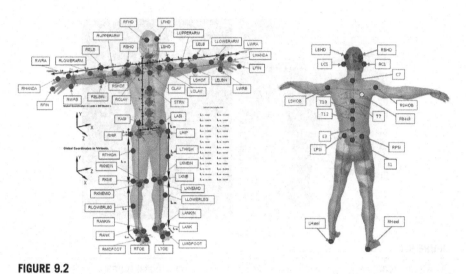

FIGURE 9.2

Santos® marker placement protocol.

the acromion and the axilla, over the medial and lateral epicondyles of the humerus, over the radial and ulnar styloid processes, and over the 2nd and 5th metacarpal heads.

Collected marker motion data are smoothed using a Butterworth filter; the cut-off frequency is usually 8 Hz, but it would depend on the frequency content of the motion for high-speed motion. The subjects are instructed to stand in a neutral position, referred to as the T-pose, which corresponds to the initial joint angles and segment locations of the skeleton. The T-pose is defined as standing with feet shoulder width apart and parallel, and arms raised parallel to the floor in the transverse plane and lateral to the body in the frontal plane. Palms face forward with the elbows maximally extended and the olecranon process pointed towards the ground.

9.3.4 Subject preparation and data collection

When a subject comes into the motion capture lab, they put on the motion capture suit, and markers are placed on their body according to the previously defined marker placement protocol. Bony landmarks are carefully located, and corresponding markers are placed accordingly to reflect these anatomical reference points (Karduna et al., 2001; Oyama et al., 2001). Figure 9.3 shows the marker placement schemes.

Subjects are first instructed to stand in the T-pose. The first trial in each motion capture was a range of motion (ROM) trial. In the ROM trial, the subject performed a series of movements that isolated each joint. Some of these movements were head flexion/extension, arm circles, hip rotations, bending

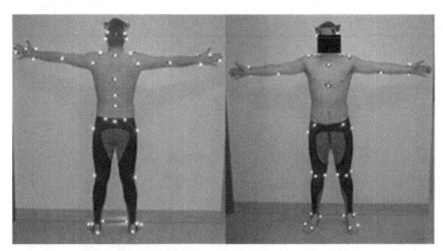

FIGURE 9.3

Marker placement schemes (T-Pose).

side-to-side and ankle rolls. Movements were designed to isolate joints or DOF in a joint, and to identify joint limits. The knee was modeled as a single-DOF hinge joint. Knee flexion during the ROM trial isolated this joint. However, the hip was modeled as a 3-DOF ball-and-socket joint, so hip flexion isolated a single DOF in the hip joint. Hip abduction and rotation were performed separately to analyze the remaining DOF in the hip.

9.4 Methods

9.4.1 Normalizing the data

The motion capture data for the throwing task are considered in this section. The time history of the joint angles for each subject shows variations in the total task time, making a direct frame-by-frame comparison between the experiment and prediction infeasible. Therefore, a specific normalized cycle time is defined based on distinguishable human positions during the motion. For example, during throwing (Figure 9.4), it is found that the initial cocked position and object release point provide two comparable positions between subjects. The throwing cycle contains all of the motion from the cocked position until the object is released from the hand. Since the motion following the release point is also important, an additional 10% of data after the release point is also considered.

With this defined cycle, all subjects start and end in the same position independent of the time elapsed between. In this case, the normalized time can be

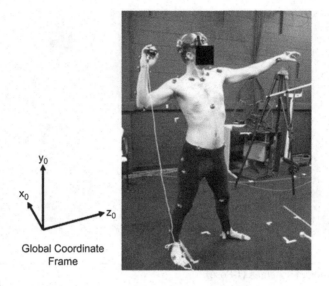

Global Coordinate
Frame

FIGURE 9.4

Subject in the initial posture of the throwing task.

used directly to compare the joint angles based on their position in the cycle. The time normalization algorithm can be seen in other studies, such as those by Matsunanga et al. (2004) and Nadzadi et al. (2003). In this work, the starting cycle time is defined as $t = 0$, and the ending (release point) time as $t = 1$.

In other words,

$$Proportion\ of\ time = \frac{(t - t_0)}{(t_n - t_0)} \tag{9.1}$$

where t_0 represents the time point at which the subject is ready to throw the object and t_n represents the time point at which the subject releases the object from their hand. This approach will be used throughout the chapter.

9.4.2 Validation methodology

The validation methodology involves testing the validity of the PD task to pass effectively and consecutively through four benchmark tests comprising two qualitative and two quantitative comparisons. In general, the benchmark tests depicted in Figure 9.5 are ordered by increasing levels of strictness of conformity and validation effort.

In the qualitative comparison stage, the first benchmark test asks observers to compare movies or pictures simultaneously played back by two avatars. One avatar uses experimental data, and the other uses predictive data. The first benchmark test is a very helpful tool that plays a significant role in capturing the differences

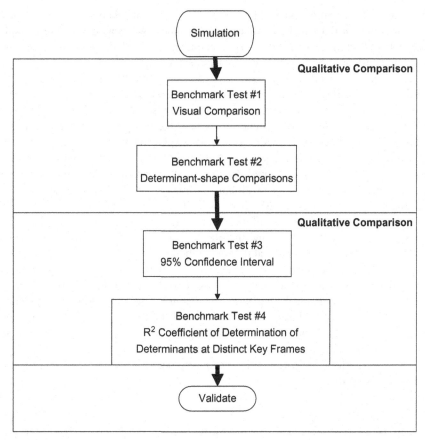

FIGURE 9.5

Flow diagram of the validation methodology showing the two comparison stages with the four benchmark tests.

in subjects' trends during complex motions where people use different strategies that cannot be captured by the quantitative statistical measures. The second benchmark involves subjectively comparing the general shape of the joint angle time histories for the experimental and predicted avatars.

In the quantitative comparison of the validation methodology, the third and fourth benchmarks quantitatively compare the determinants' time histories based on statistical metrics. The quantitative comparison stage comprises rigorous statistical comparisons between the magnitude of the predicted and normal determinants. The first benchmark of this stage tests if the predicted time history magnitude of the determinants follows the magnitude of the mean of the normal subjects and falls within the 95th percentile interval of confidence. The second and final benchmark of this stage is designed to statistically determine, using the coefficient of determination R^2, the degree of correlation between the predicted and experimental data at the distinctive key frames. All benchmarks are considered valuable because substantial effort could be saved if the predicted motions could be configured during the early benchmark tests.

The validation methodology performs each benchmark test by comparing the predicted motions against the average motions obtained by performing experiments for the same task. One critical issue that needs to be considered in the validation process is the inconsistency between the experimental data and the predicted data spaces. The experimental 3D displacement data are expressed in the Cartesian space, while the predicted dynamic tasks are expressed in the joint space. Therefore, the experimental data are first transformed from the Cartesian space to the joint space using a global optimization-based IK scheme (Lu and O'connor, 1999; Xiang et al., 2011).

Due to the large amount of information in the resulting motion, the validation process becomes cumbersome, and therefore it only considers the determinants of the motion. Furthermore, the validation methodology selects a more restrictive subspace of these determinants at distinct key frames. The distinct key frames represent the magnitude of the determinants at well-defined times in the determinants' time history. For a lifting task, for example, the magnitudes of the knee angle at 0%, 20%, 40%, 60%, 80%, and 100% of the lifting height can be used as distinct key frames. In addition to validating the kinematics of the motion, the validation methodology also checks the kinetics of the motion.

Two tasks will be validated in this chapter to demonstrate the method, process, and accuracy of PD.

9.5 Validation of predictive walking task
9.5.1 Walking task description

Following the ROM, the subjects performed the normal walking trials. The normal walking trial was set up so that the subjects walked continuously back and forth across the laboratory. The first and final steps of each pass were considered acceleration and deceleration, and the middle stride was analyzed as a steady-state gait cycle for normal forward walking.

The normal walking trial was set up so that the subjects walked forward at a comfortable speed, stopped, and walked backwards to return to the starting position. On average, the subjects took 4−5 steps forward depending on preferred step length.

The trials were repeated to ensure that the results were usable and to examine the consistency of the velocities chosen by the subjects. In addition, every trial began and ended with the subject in the same position—the T-pose described earlier in this chapter. The reflective markers seen on the subject in Figure 9.6 correspond to the Santos® marker placement protocol.

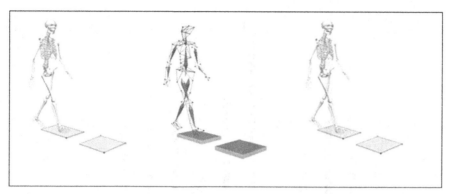

FIGURE 9.6

A picture of a subject's skeleton motion capture during walking, and the ground force plates.

9.5.2 Walking determinants

Based on the literature and a fair understanding of human gait, six determinants were chosen to characterize walking (Saunders et al., 1953). These walking determinants (WD) represent major DOF, or a combination of DOF, that play considerable roles in the walking process. Walking determinants consist of the lower extremities and the pelvic motion of the human, and include hip flexion/extension, knee flexion/extension, ankle plantar/dorsiflexion, pelvic tilt, pelvic rotation, and lateral pelvic displacement. With the exception of lateral pelvic displacement, each of the gait determinants is the time history of a joint angle. Pelvic motion, which includes pelvic tilt, pelvic rotation, and lateral pelvic displacement, is not consistently identified in the literature; however, it will be included in this work because of its significant involvement in normal human walking.

9.5.3 Participants

Nine healthy males and one healthy female were used for the validation of normal walking. The subjects had no history of musculoskeletal problems and were reasonably fit. Their participation was voluntary, and a written informed consent, as approved by the University of Iowa Institutional Review Board, was obtained prior to testing. The height of the subject population was 176 ± 7 cm, the weight was 67 ± 7 kg, and the age was 27 ± 7 years.

9.5.4 Results

9.5.4.1 Qualitative comparison

For the first benchmark (Figure 9.5), 13 participants were used to evaluate five videos of animated avatars. Four avatars were animated using experimental data,

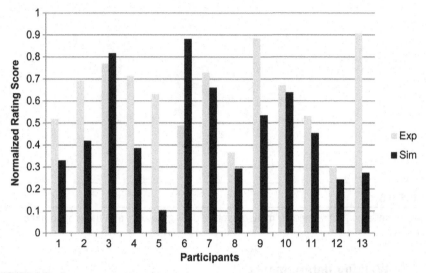

FIGURE 9.7

Normalized rating scores of 13 participants evaluating the predicted and experimental walking.

and one was animated using predicted data. Figure 9.7 shows a comparison of the rating scores. The p-value of the Wilcoxon Signed Rank Test was 0.0215. The exact p-value (p-value = 0.133) calculated from binominal distribution indicates that the percentage of people who thought the simulated walking was normal was not significantly less than 50%.

For the second benchmark, Figure 9.8A shows that the predicted determinant-curve shapes for walking have general shapes that closely follow the individual 10 subjects. Figure 9.8B demonstrates the coupling strength between one of the predicted determinants—hip flexion, for example—with other WD. As shown in Figure 9.8B, the predicted hip determinant almost captured the strong coupling between the determinants of the natural subjects.

Figure 9.8A demonstrates a comparison between the velocity of the walking determinants of the 10 subjects and the model. As shown in the figure, the shape of the predicted model velocity is, in general terms, similar to that of the subjects. However, there are fewer local fluctuations in the predicted WD.

Calculated acceleration (Figure 9.9B) has shown behavior similar to that of the velocity depicted in Figure 9.9A; however, more local fluctuations appeared in the experimental curves as a result of the finite difference calculations. For kinetics, Figure 9.10 shows the experimental and predicted vertical and forward ground reaction forces during walking. As indicated in the figure, the predicted model showed comparable characteristics to those of the experimental data.

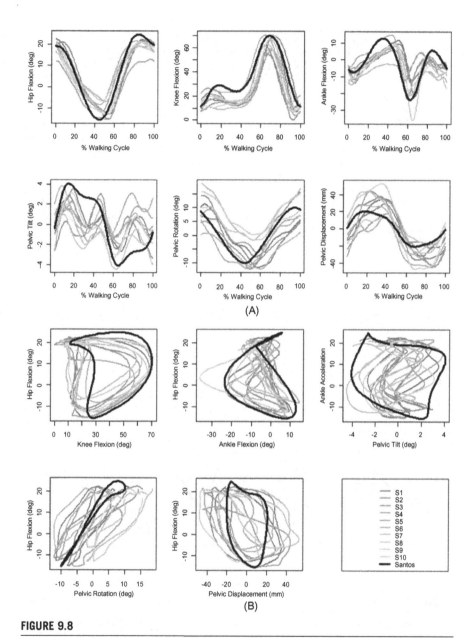

FIGURE 9.8

(A) Comparisons of the joint profiles (of the determinants) obtained from validation experiments versus those predicted; (B) coupling strength between one of the predicted determinants (hip flexion) and the rest of the determinants.

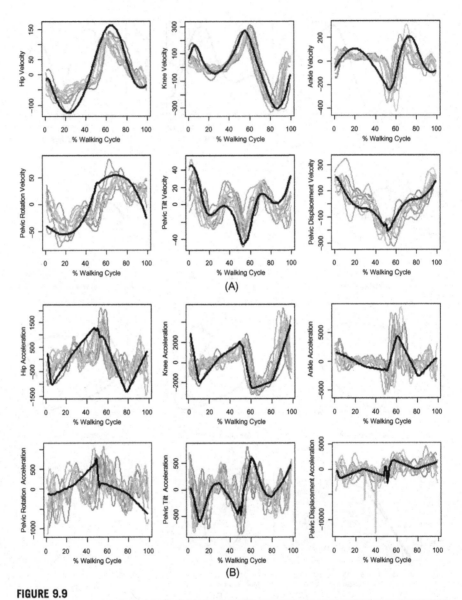

FIGURE 9.9

(A) Comparisons of joint velocity profiles versus those obtained from the validation experiments; (B) comparisons of joint acceleration profiles versus those obtained from the validation experiments.

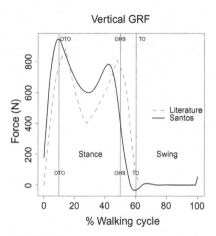 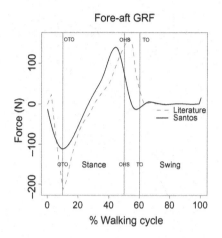

FIGURE 9.10

Ground reaction forces (GRF) in the vertical and fore-aft directions during the walking cycle of Santos® (solid line) and those in the literature (Stansfield et al., 2006) (dashed line).

9.5.4.2 Quantitative comparison

The comparison between the experimental and predicted WD in terms of the interval of confidence is presented in Figure 9.11. The model determinants showed significant correlation and agreement with the human subjects, by being inside the interval of confidence and following the trajectories of the subjects' determinants. Figure 9.12 shows the selected key frames for the walking task, while Figure 9.13 depicts the R^2 plot for the key frames in Figure 9.12. The diamond shape represents the relationship between the simulation data and the average experimental data of the hip flexion of 10 subjects. Other graphs represent the relation for the rest of the determinants.

9.5.4.3 Discussion

The results for the first benchmark test showed that 30% of the participants who watched five videos of the predicted and experimental models thought the simulation video looks normal. It should be noted here that some participants gave low scores for some of the experimentally driven avatars, indicating that people have different levels of visual perception in detecting abnormalities in the way avatars are moving. The results of the second benchmark test show significant similarity between the shapes of the simulated and human WD. The predicted model even shows some potential to capture the strong coupling between the WD.

For the quantitative benchmark tests, and in general terms, the results have shown that the model's hip flexion/extension time history reaches relatively greater maximum and minimum peaks than the mean of the subject population; however, it is still considered within normal behavior. Knee flexion/extension lies

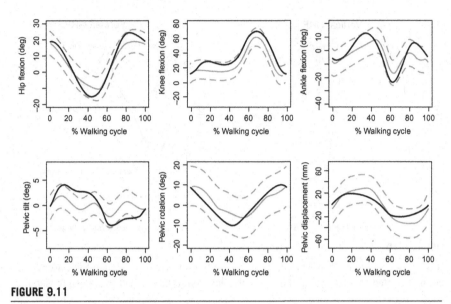

FIGURE 9.11

The 95% confidence intervals are shown in dashed lines for the walking determinants. The experimental subject's mean is shown in gray and Santos®'s determinants are shown in black.

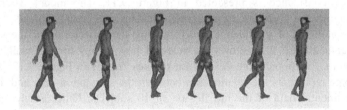

FIGURE 9.12

Selected key frames of the walking task.

within the region; however, the simulation curve, while shifted vertically, is still within the interval of confidence. There is a lesser degree of correlation between the ankle dorsi-flexion and normal human motion, specifically in the second third of the cycle. The general trend line of the pelvic tilt seems to agree with the experimental curves. The pelvic tilt falls outside the 95% confidence interval from 35% to 50% of the gait cycle when the ANOVA test shows that $P < 0.001$. The pelvic rotation of the predicted model is well correlated to that of the human subjects. In terms of lateral pelvic displacement, the simulation provides a relatively normal curve, but the peak values occur at different percentages of the gait cycle than were seen in the subject population.

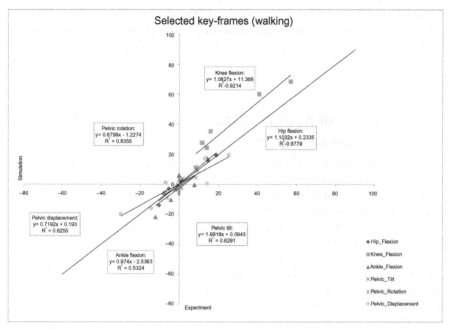

FIGURE 9.13

R^2 plot for the six walking determinants for the selected key frames (bottom). The vertical axis (in degrees) stands for simulation data (Santos®), and the horizontal axis (in degrees) corresponds to the average experimental data.

For the key frames, Figure 9.13 demonstrates a good correlation for the hip flexion, as indicated by the value of R^2, which exceeds 0.9. Similar behaviors with less correlation can be seen for other determinants. However, the ankle dorsi-flexion has shown relatively poor correlation, with R^2 less than 0.53. For the GRF, Figure 9.10 demonstrates a good agreement between the two curves in the vertical and the fore-aft directions. This gives the model significant value and potential for calculating realistic values for the joint torque of the different human's joints.

The determinant's velocity for the predicted model has shown similar characteristics to those of the subject population. However, the predicted model could not capture the higher-frequency components, which are represented by the local fluctuation in the velocity curves. Normally, natural human motion is relatively smooth, and the local fluctuations, shown in Figure 9.9A, are anomalies that may represent some type of numerical noise as a result of the finite differences calculation, or because of local movement of reflective markers due to skin motion (Garling et al., 2007; Lucchetti et al., 1998). A low-pass filter may be used to smooth out the experimental data. What has been shown and applied to velocity seems to be applicable to the acceleration, as shown in Figure 9.9B.

The results, in general, have shown that PD models can be relied upon to realistically predict human walking motion and GRF. They also indicate the potential of future contributions of these models to the area of human biomechanics.

9.6 Validation of box-lifting task

9.6.1 Lifting task description

Subjects were instructed to stand in a neutral position, referred to as the T-pose (Figure 9.3), which corresponds to the initial joint angles and segment locations of the skeleton. It is well known that subjects may use different lifting strategies depending on their strength and their perception of the load (Bartlett et al., 2007; Li and Zhang, 2009). In this work, participants were aware of the weight of the box (20 lb for males and 15 lb for females) and were shown proper material-lifting strategies to avoid any unexpected harmful strategies. They were then instructed to lift a box from the standing surface to shoulder height in their most natural or comfortable way (Burgess-Limerick et al., 1995). Adequate warm-up and rest time was allotted. Figure 9.14 shows the initial (left picture) and final (right picture) lifting postures.

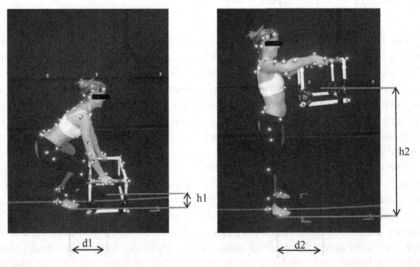

FIGURE 9.14

Initial and final lifting postures and the input parameters for the predicted box-lifting task (d1, d2, h1, and h2).

9.6.2 Box-lifting determinants
Based on the literature, six determinants—torso flexion [1], shoulder flexion [2], elbow flexion [2,3], hip flexion [2,3,4], knee flexion [2,3,4], and ankle flexion [2,3,4])—are identified in this work to determine if the PD model's (Santos®'s) box-lifting motion correlates to the real-world observations.

9.6.3 Participants
Twelve subjects participated in the subjective evaluation of the predicted motion, and another 12 subjects participated in the testing of a whole-body motion task. The first 12 subjects were asked to watch movies of two avatars, one driven by simulation data and one driven by experimental data, and then to evaluate the avatars' motion normality. The second 12 subjects (8 healthy males and 4 healthy females) were involved in the experimentation of the materials-lifting task. The subjects had no history of musculoskeletal problems and were reasonably fit. Their participation was voluntary, and a written informed consent, as approved by the University of Iowa Institutional Review Board, was obtained prior to testing. The height of the subject population for the materials-lifting task was 175.6 ± 11.5 cm, the weight was 73.2 ± 10.7 kg, and the age was 21.8 ± 4.3 yrs.

9.6.4 Results
9.6.4.1 Qualitative comparison
Twelve subjects participated in the first benchmark test to observe and evaluate two videos of animated avatars. One avatar was animated using experimental data, and the other was animated using predicted data. The participants were asked to report their scores based on what they consider a natural human lifting motion. Figure 9.15 shows a comparison of the rating scores for both avatars. The coefficient of correlation (r) between the experimental and simulation rate was 0.692729.

For the second benchmark, Figure 9.16A shows that the predicted determinant-curve shapes, except for the ankle flexion, for lifting during the lifting cycle have general shapes that closely followed the 12 individual subjects. Figure 9.16B demonstrates the coupling strength between one of the predicted determinants—shoulder flexion, for example—and other lifting determinants. As shown in the figure, the predicted shoulder flexion determinant followed to some extent the trend of coupling between the determinants of the natural subjects. Again, the predicted ankle flexion showed weak correlation with the experiments.

Figure 9.17A shows a comparison between the velocity of the lifting determinants of the 12 subjects and the model. As shown in the figure, the shape of the predicted model velocity is similar to that of the subjects for the torso and hip. However, there are fewer local fluctuations in the predicted lifting determinants. The shape of the rest of the determinants was somehow different. The calculated

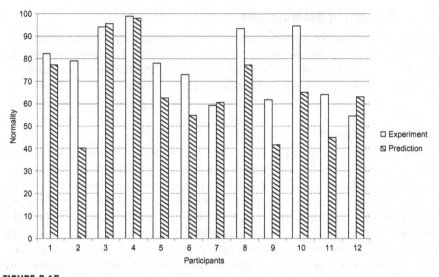

FIGURE 9.15

Normalized rating scores of 12 participants evaluating the predicted and experimental box-lifting task.

acceleration (Figure 9.17B) shows behavior similar to that of the velocity; however, more local fluctuations appeared in the experimental curves.

For the two important joints for this task (hip and torso), the calculated velocity and acceleration profiles are shown as compared with values obtained from the experiments (Figure 9.17). It is shown that the PD method yields adequate results and is indeed a rigorous approach to predicting the motion.

For kinetics, Figure 9.18 shows the experimental and predicted vertical and forward GRF during lifting. As indicated, the predicted model showed behaviors similar to those of the experimental data but was not able to capture the initial characteristics of the lifting cycle.

9.6.4.2 Quantitative comparison

The comparison between the experimental and the predicted lifting determinants in terms of the interval of confidence is presented in Figure 9.19. The experimental data for each subject represents the average of two lifting cycles. The model determinants show weak correlation for the ankle and the shoulder flexion; however, they show reasonable correlation and agreement with the human subjects for the rest of the determinants, by being inside the interval of confidence and following the trajectories of the subjects' determinants. Figure 9.20 depicts the R^2 plot for the six lifting determinants for the distinctive key frames (0% height, 20% height, and 40% height). The solid black circle in Figure 9.20 represents the relationship between the simulation data and the average experimental data of the hip

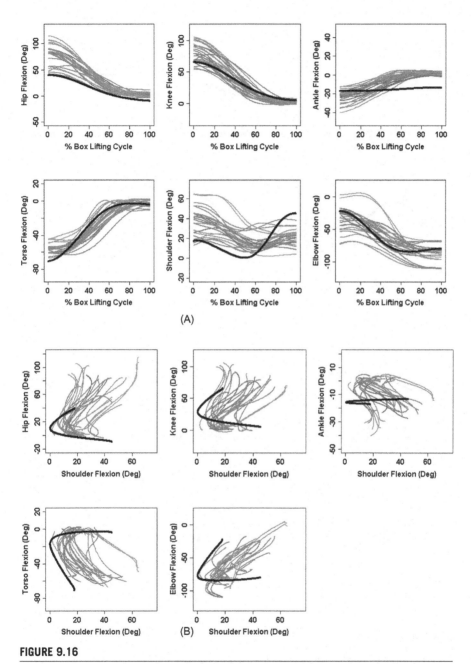

FIGURE 9.16

(A) Comparisons of the joint profiles (of the determinants) obtained from validation experiments versus those predicted; (B) coupling strength between one of the predicted determinants (shoulder flexion) and the rest of the determinants.

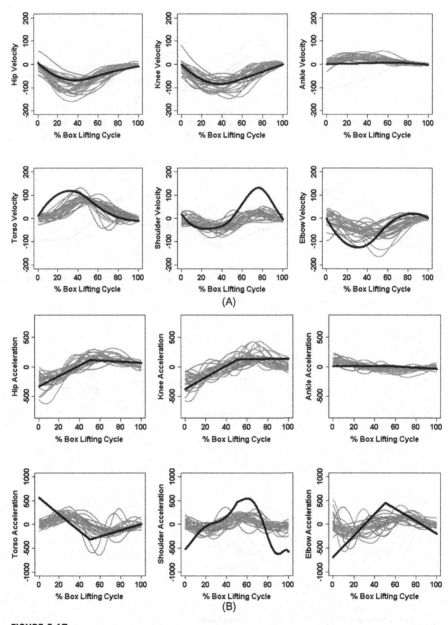

FIGURE 9.17

(A) Comparisons of joint velocity profiles versus those obtained from the validation experiments; (B) comparisons of joint acceleration profiles versus those obtained from the validation experiments.

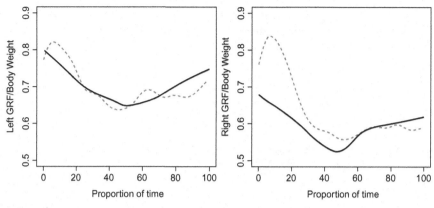

FIGURE 9.18

Ground reaction forces in the vertical direction during the box-lifting cycle (dashed gray curve represents experimental results, and solid dark curve represents the predicted results).

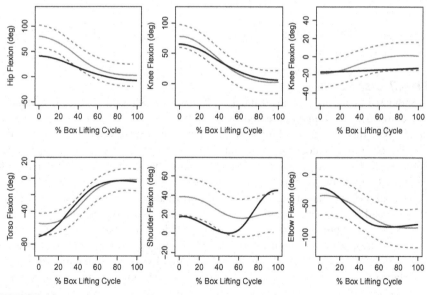

FIGURE 9.19

Interval of confidence for each critical parameter.

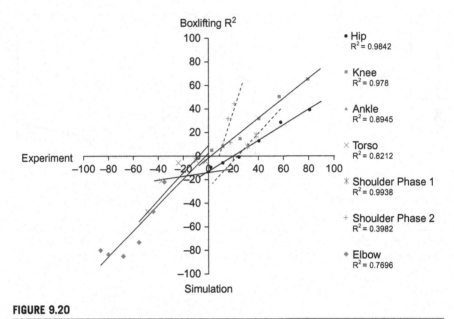

FIGURE 9.20

The R^2 values for experiment versus predictive simulation.

flexion of 12 subjects. Other symbols in Figure 9.20 represent the relation for the rest of the determinants.

9.6.4.3 Discussion

As indicated in Figure 9.5, the validation framework is based on four benchmark tests comprising two qualitative and two quantitative benchmark tests. The results for the first qualitative benchmark test (which demonstrates subjective comparisons by observing videos of the simulated and experimental models shown in Figure 9.15) have shown a reasonable correlation between the experimental and the predictive models with an r-value of 0.692729.

The subjective evaluation may not give information as accurate or specific as that obtained by the objective evaluation, but it can give some expectations to the modeler that the objective measure cannot provide. For example, the subjective measure can present how realistically the whole body is moving during the task. It also presents images that may help in finding or localizing unnatural behaviors at certain joints that could be hard to observe by just looking at the data coming from the objective evaluation. The authors found that the subjective assessment was very useful for the modeler, especially when the motion is far from natural, or the strategy of the task is different from what the model predicts. While the subjective assessment gives general information, it can still point out specific problems that can inform the quantitative assessment. For example, major differences in the whole-body motion will make the modeler think about a different objective

0% height   20% height   40% height   60% height   80% height   100% height

FIGURE 9.21

Predicted model showing human posture during the selected key frames.

FIGURE 9.22

Experimental model showing human posture during the selected key frames.

function. Or, a major problem at a joint will make the modeler add, delete, or relax some constraints at these joints.

As shown in Figure 9.15, some participants gave low scores to the experimentally driven avatar. One reason for this poor evaluation is related to the difficulties associated with the level of visual perceptions and deceptions of humans when they watch moving objects in the virtual world. Another reason for this rating could be the way the avatar looks. As shown in Figures 9.21 and 9.22, the avatars used for the animation have very detailed realistic skin and human-like features. These features may distract the observers and cause them to unintentionally check other irrelevant avatar attributes. For example, the participants were asked to focus on and evaluate the lifting determinants, but some participants commented on the way the avatar's head was moving. In general, the first benchmark test provided significant information about the acceptance of the model; however, it provides limited information regarding the accuracy of the determinants. This is because the model has a large DOF with strong coupling between the determinants, which makes it very hard for normal eyes to differentiate between the predicted and experimental determinants.

The results of the second qualitative benchmark test have shown some similarity between the shapes of the predicted and human lifting determinants as shown in Figure 9.16A and Figure 9.16B. However, the model showed poor performance in following the shape of the ankle and shoulder flexion during the lifting cycle. The predicted model showed some potential for capturing the coupling between the lifting determinants as shown in Figure 9.16B, with some poor performance in capturing the coupling between the shoulder and the ankle. The importance of this benchmark test is to detect major discrepancies, if there are any. Additionally, this test can identify any abnormal characteristics in the shape of the determinants' time history during the lifting cycle.

In terms of the quantitative comparison in the third benchmark test, the predicted lifting motion showed a reasonable correlation with the experimental data (Figure 9.18), where all determinants, except the shoulder flexion-extension and the ankle flexion angles, stayed inside the interval of confidence at all times and, notably, followed the mean of the subjects. The shoulder flexion-extension angle showed acceptable correlation during the first half of the lifting cycle, but poorly represents human characteristics during the second half of the motion cycle. These discrepancies could be related to model accuracy, in that the experimental data was collected from a real human with muscle and skin movement, while the model is based on a rigid-body dynamics assumption.

The discrepancies may also be due in part to the absence of necessary constraints on the complex shoulder motion in the model and in part to the difficulties associated with computing accurate shoulder joints from the experimental data. The discrepancies could also be attributed to other parameters, such as the complex interaction of the shoulder-clavicle complex motion and the motion of the wrists when the box is extended above chest level while keeping the feet on the floor. Ankle flexion showed behaviors similar to those of shoulder flexion.

Unlike other human motion tasks that require a single common strategy, such as walking, the box-lifting task can be conducted using different strategies; therefore, the whole motion looks natural, but the determinants' shapes could be affected by the coupling and the speed associated with the strategy being used. As shown in Figure 9.21, the model inadequately captured the early stages of the lifting process, which could involve different speeds and different lifting strategies; still, the model succeeded in capturing the following steady-state phase for both legs. It should be noted here that the GRF for box lifting were based on one subject and may be not sufficient for use in model kinetics validation due to the uncertainty in the lifting strategy and the possibility of the subjects leaning laterally. Therefore, the reaction forces at the hands should also be measured and used in the comparison. However, in this work, we considered symmetrical lifting and assumed equal distribution of the forces on the hands and legs.

For the key frames in the fourth and last quantitative benchmark test, Figure 9.22 demonstrates the correlation between the experimental and simulated determinants with R^2 extended from 0.72 to 0.98 for most determinants, with the exception of the shoulder flexion.

As mentioned earlier, the shoulder performed adequately during the first half of the lifting cycle (Figure 9.18); however, the coupling between the shoulder-clavicle motion, the stretching in the trunk, the twisting in the wrists, and the additional motion from the lower extremity may have affected the results. For these reasons, the graph for the shoulder extension, shown in dashed lines in Figure 9.19, has been divided into two segments representing the first and second halves of the lifting cycle. Interestingly, in the first qualitative benchmark test, most subjects did not observe any abnormality in the shoulder motion because it is too local.

9.7 Feedback to the simulation

Validation is not only necessary but essential to improving the simulation. When there is a major difference between the predicted motion and human motion, it is critical that the PD formulation be refined. This is accomplished by using different objective functions in the optimization schemes. If unrealistic motion has occurred at a certain joint, it may become clear that a physical constraint is missing or that an additional objective function is needed to render the motion more natural or realistic.

For example, for the box-lifting task, the subjective evaluation showed that the whole-body motion of the predicted model (Figure 9.20) looks different than the whole-body motion of the experimental model (Figure 9.21) for the key frames 0% height, 20% height, and 40% height, which could be related to the lifting strategy. In this case, the objective evaluation (Figures 9.18 and 9.19) may not have the capacity or the information to relate that to the change in the strategy, because the discrepancies in the motion may show in parts of some determinants, which could be considered as a deficiency in the model. This is very clear in the objective evaluation of the determinants (Figure 9.18), where all determinants were inside the interval of confidence except the hip flexion and shoulder flexion, which were outside the interval of confidence for the first 40% of the cycle. This is a good example for showing the significance of having the objective and subjective evaluations in the validation framework.

9.8 Concluding remarks

This chapter has presented a framework to validate the results of PD. The framework was applied, as an example, to a 55-DOF predictive computer human model performing two of the most important common tasks: walking and box lifting.

Validation of human motion is not an easy process; however, it is an essential step of formulation development. There are two purposes for validation: (a) to ascertain that the formulation for predicting a human motion is correct, and more importantly, (b) to provide feedback to the modeler, and subsequently the model, in order to change the formulation to a higher level of fidelity.

In the PD formulation, there are many ways to change the formulation effectively and with relative ease. Changing the cost functions that drive the motion is probably the most direct and effective method for affecting the motion. Obviously, more insight into the biology, the task, and the output will provide better control and input into the model. Adding, removing, tightening, and relaxing constraints will also affect change.

The validation framework shows that the predictive model considered in this work can predict, to a certain level, human motion during walking and box lifting. However, the model needs additional work to capture the characteristics of natural human motion. The proposed validation method was able to localize the problems in the model and showed that it has difficulty capturing some aspects of the task dynamics, such as the initial GRF profile and some of the task kinematics like ankle and shoulder flexion profiles. The results from this validation framework were used to locate specific abnormal characteristics in the motion determinants and to provide feedback to the developers for further refinement of the predicted lifting task.

References

Abdoli-Eramaki, M., Stevenson, J.M., Agnew, M.J., Kamalzadeh, A., 2009. Comparison of 3D dynamic virtual model to link segment model for estimation of net L4/L5 reaction moments during lifting. Comput. Methods Biomech. Biomed. Eng. 12 (2), 227–237.

Baliunas, A.J., Hurwitz, D.E., Ryals, A.B., Karrar, A., Case, J.P., Block, J.A., et al., 2002. Increased knee joint loads during walking are present in subjects with knee osteoarthritis. Osteoarthr. Cartil. 10 (7), 573.

Bartlett, D., Li, K., Zhang, X., 2007. A relation between dynamic strength and manual materials handling strategy affected by knowledge of strength. Hum. Factors 49, 438–446.

Blankevoort, L., Huiskes, R., 1996. Validation of a three-dimensional model of the knee. J. Biomech. 29 (7), 955–961.

Burgess-Limerick, R., Abernethy, B., Neal, R.J., Kippers, V., 1995. Self-selected manual lifting technique: functional consequences of the interjoint coordination. Hum. Factors 3 (2), 395–411.

Chaffin, D.B., 2002. On simulating human reach motions for ergonomic analysis. Hum. Factor Ergon. Man. 12 (3), 235–247.

Chaffin, D.B., 2009. The evolving role of biomechanics in prevention of overexertion injuries. Ergonomics 52 (1), 3–14.

Dubowsky, S.R., Rasmussen, J., Sisto, S.A., Langrana, N.A., 2008. Validation of a musculoskeletal model of wheelchair propulsion and its application to minimizing shoulder joint forces. J. Biomech. 41 (14), 2981–2988.

Garling, E.H., Kaptein, B.L., Mertens, B., Barendregt, W., Veeger, H.E.J., Neilissen, R.G.H.H., et al., 2007. Soft-tissue artefact assessment during step-up using fluoroscopy and skin-mounted markers. J. Biomech. 40, S18–S24.

Karduna, A.R., McClure, P.W., Michener, L.A., Sennett, B., 2001. Dynamic measurements of three-dimensional scapular kinematics: a validation study. J. Biomech. Eng. 123 (2), 184−190.

Lucchetti, L., Cappozzo, A., Cappello, A., Della Croce, U., 1998. Skin movement artefact assessment and compensation in the estimation of knee joint kinematics. J. Biomech. 31, 977−984.

Li, K., Zhang, X., 2009. Can relative strength between the back and knees differentiate lifting strategy? Human Factors 51 (6), 785−796.

Lu, T.W., O'connor, J.J., 1999. Bone position estimation from skin marker co-ordinates using global optimisation with joint constraints. J. Biomech. 32 (2), 129.

Marras, W.S., Sommerich, C.M., 1991. A three-dimensional motion model of loads on the lumbar spine: II. Model validation. Human Factors 33 (2), 139−149.

Matsunaga, N., Okimura, Y., Kawaji, S., 2004. Kinematic analysis of human lifting movement for biped robot control, 2004. AMC '04. The 8th IEEE International Workshop on Advanced Motion Control. pp. 421−426.

Nadzadi, M.E., Pedersen, D.R., Yack, H.J., Callaghan, J.J., Brown, T.D., 2003. Kinematics, kinetics, and finite element analysis of commonplace maneuvers at risk for total hip dislocation. J. Biomech. 36 (4), 577−591.

Oyama, E., Agah, A., MacDorman, K.F., Maeda, T., Tachi, S., 2001. A modulator neural network architecture for inverse kinematics model learning. Neurocomputing 38, 797−805.

Rabuffetti, M., Baroni, G., 1999. Validation protocol of models for centre of mass estimation. J. Biomech. 32 (6), 609−613.

Robert, T., Cheze, L., Dumas, R., Verriest, J.-P., 2007. Validation of net joint loads calculated by inverse dynamics in case of complex movements: application to balance recovery movements. J. Biomech. 40 (11), 2450−2456.

Saunders, J.B.D.M., Inman, V.T., Eberhart, H.D., 1953. The major determinants in normal and pathological gait. J. Bone Joint Surg. Am. 35-A (3), 543−558.

Xiang, Y., Rahmatalla, S., Arora, J.S., Abdel-Malek, K., 2011. Enhanced optimization-based inverse kinematics methodology considering joint discomfort. Int. J. Human Factors Model. Simulat. 2, 111−126.

Concluding Remarks

10

*I think and think for months and years. Ninety-nine times, the **conclusion** is false. The hundredth time I am right.*
Albert Einstein

This book has introduced a new method for predicting and analyzing human motion. While it is not perfect, it is a broadly applicable formulation for many tasks that humans perform. Predictive dynamics (PD) has been shown to be a general systematic method for predicting human motion. In this chapter, we summarize the most important aspects of this method, its applications, and future research endeavors.

10.1 Benefits of predictive dynamics

We enumerate below some of the benefits of the PD method.

10.1.1 Using the Denavit–Hartenberg (DH) method is effective in modeling human kinematics

The DH parameterization method is an effective and powerful approach to modeling human segments, embedding the coordinate systems, and creating the four generalized coordinates that relate one frame to another. Indeed, the DH method provides for a systemic approach to modeling the kinematics.

Because the DH representation method is so consistent, it is a straightforward process to represent the underlying kinematic skeleton and use it as a reference to attach rigid objects to. Since each body is defined with respect to a coordinate system, it was also possible to define additional rigid bodies, such as clothing items, loads, helmets, and any other object that affects human motion, to that frame.

Additionally, because the DH parameters are consistent and flexible in their ability to be programmed into a code, changes in body stature and size (anthropometry) and changes in joint limits can be readily implemented. As a result, PD can analyze an individual of a specific weight, height, body type, and strength and fatigue capability. All of these characteristics are input parameters in the PD environment.

10.1.2 Predictive dynamics solves dynamics without integration

This is perhaps the most significant benefit of PD. Multi-body dynamics typically attempts to use numerical integration to determine the motion of a system of rigid bodies that has been subjected to external forces or is in motion. However, multi-body dynamics methods cannot address the prediction of human motion for the following reasons:

a. Multi-body dynamics cannot handle high-DOF models having equality and inequality constraints on the motion, such as those of the human body.
b. The algebraic-differential equations that are formed by the human system are too complex and are highly nonlinear.
c. Unlike PD, multi-body dynamics does not consider the natural behavior of the motion. The PD optimization formulation, on the other hand, uses objective functions to drive the motion.

10.1.3 Predictive dynamics renders natural motion

The essence of the PD method is its ability to render natural motions, similar to those of humans. It considers the human body to have a set of cost functions that drive the motion. This is an essential difference from traditional inverse kinematics methods, which yield postures and motions that cannot be performed realistically by humans. Predictive dynamics exhibits natural motion because the motion is constrained by the joint range of motion for all joints, the laws of physics, and any physical constraints that may exist. More importantly, PD attempts to find solutions that obey the dynamic strength surface for each joint. The result is natural motion that looks realistic for a person of a certain stature and strength.

10.1.4 Predictive dynamics induces natural behavior

The use of one or more cost functions, depending on the task, has been shown to yield different behaviors. It was also shown that using multi-objective optimization can induce certain behaviors typically performed by humans. Whether it is energy, jerk, discomfort, effort, or torque, it is obvious that humans try to minimize or maximize some quantity. This, we believe, also develops as humans gain experience, grow wiser, and become adept at performing various tasks.

10.1.5 Predictive dynamics admits cause and effect

A human simulator can be tried, loaded, and pushed to obtain a reaction. The user of a human simulator that is developed using PD can experience cause and effect. As shown in the box-lifting example, a higher load will cause substantially higher torques. An enormous load will cause the human simulation system to respond with "I cannot do this." The PD method will simply not converge and will be stopped.

Cause and effect is one of the most important aspects of a simulator. It is what differentiates an analysis system from an animation system. In gait analysis, for example, the analyst obtains data for the physical gait from the person and then employs software tools to analyze the motion and determine its performance. However, if this same analyst wanted to examine a different condition, the analyst would have to return to the laboratory, conduct another experiment, and perform another analysis. With a human simulator, changing the conditions and determining the effects of, or response to, that change can occur immediately in the software without the need to do additional experiments.

10.1.6 **Predictive dynamics uses joint space, not muscle space**

One of the most significant contributions from our work over the past 10 years has been the realization that modeling and working with joint space is effective. Instead of addressing the resolution of forces into every muscle or muscle group, we focused our attention on generalized variables at the joint level—namely, rotation angle, angular velocity, angular acceleration, and torque. Because each joint is represented by one or more DOF, each DOF is systematically associated with a set of generalized variables.

Predicting motion with PD was then simplified to focus on determining these generalized variables at each DOF. Once the predicted motion is determined, these variables are calculated as a discretized curve (a B-spline) that governs the shape/velocity/acceleration of the motion, but also with the capability to calculate torques for every DOF and reaction forces where needed. It is a fast, efficient, reliable, and robust approach to predicting human motion.

Recovery of muscle forces, if needed, can be accomplished by first combining the torques for every DOF at the joint into a resultant component, then resolving the resultant component torques into muscle action forces.

10.1.7 **Predictive dynamics uses dynamic strength surfaces**

Because we have used joints as the key ingredient in our basic formulation for PD, it is critical to identify strength requirements and, even more importantly, strength limits. We have shown that answering the questions "What is the limit of this person's ability?" and "Why can't Santos® do this task?" requires a deep understanding of joint strength limits.

As a result, over the past 10 years we have conducted a substantial number of experiments to measure the joint-level strength, focusing particularly on torque-angle (isometric) and/or torque-velocity (isokinetic) data, also called dynamic strength surfaces.

These surfaces were then modeled using various strategies for implementation into the code. It was shown that the PD optimization-based formulation uses these torque surfaces as constraints in guiding the human model to accomplish a task. The ideas of modeling by using joints, limiting their strengths, and numerically optimizing them to drive natural motion fits together well and provides for a complete methodology.

10.1.8 The PD validation process is effective

A cornerstone of any simulation formulation is its ability to be validated and verified. It has been shown that validation can be accomplished on all PD tasks, albeit with varying levels of accuracy, using a number of methods. The keys to the validation strategy are the determinants of motion that were defined earlier. We do believe, however, that technologies for motion capture and analysis will continue to evolve very rapidly and that more effective validation methods can be employed. Nevertheless, validation is only successful when the validation data feed back into the model and improve its fidelity. Predictive dynamics allows for both validation and feedback into the model.

10.2 Applications

There are many applications for the use of digital human models in general, specifically the use of PD as a method for predicting human motion and understanding cause and effect.

10.2.1 Ergonomics

Consider an operator in a manufacturing plant who has to maneuver a large and heavy object into a machine. Similarly, consider another operator who works on an assembly line inserting a fitting with a certain force into the machine's assembly. Both operators will most likely acquire musculoskeletal injuries after some time. These injuries are typically not due to static loads but rather to dynamic loads and repetitive tasks. Predictive dynamics has provided a systemic method for simulating and analyzing these tasks before the task is designed to assess its feasibility.

10.2.2 Simulating an injury or a disability

Changing lower or upper joint limit constraints in conjunction with strength constraints at a single joint can simulate a disability and will cause the model to respond differently.

Exoskeletons and wearable devices to enhance human power or simply to aid in a disability are expected to make a significant impact on our society. Whether it is assisting the older population with daily activities such as walking, or using exoskeletons to replace a particular function for people with disabilities, these devices are expected to provide ample help. Design and testing of these devices requires a fundamental understanding of their effect on a human's biomechanics.

10.2.3 Sports biomechanics and kinesiology

Predictive dynamics can also be used to obtain a greater understanding of the performance of athletes. Whether the task is a golf swing or a baseball pitch, PD can be used to quickly model it, accurately analyze it, and, most importantly, simulate the various conditions under which it is executed. Imagine the scenario of changing an individual's swing angle and foot location and then executing the simulation to see which scenario yields better results for that individual. Having the capability to simulate a sports activity to optimize for the performance has significant implications on how athletes train.

10.2.4 Human performance

Predictive dynamics can be used to simulate and analyze human performance. Consider a task where a soldier has to climb up a set of stairs to accomplish a mission while carrying a heavy load. These heavy loads continue to increase, whether it is backpack contents or tools that are being added onto the soldier. Bulk added onto the soldier, reductions in flexibility or range of motion, and additional weight being carried are some of the factors that are readily considered in the PD formulation. This helps the designer of new equipment to assess and better understand the influence of such changes on the soldier's performance.

Predictive dynamics has also been implemented to address performance issues in assembly line workers in manufacturing plants or construction workers. It can be used in any situation where simulation is needed prior to subjecting a human being to unknown conditions.

10.2.5 Testing equipment, digital prototyping, human systems integration

Dynamic tasks that humans perform on a daily basis can be tested and analyzed using PD. These tasks vary in complexity and in their need for accuracy.

Digital prototyping is now used consistently by the majority of large companies all over the world. Instead of physical prototyping, computational power is used to run virtual experiments in the digital world. It is used to assess many aspects of engineering, including aerodynamics, thermodynamics, mechanical stresses, and many others. Digital prototyping has saved companies time and resources, and PD allows humans to be involved in the digital prototyping process.

Here are some examples of cases where PD has been effectively used to predict cause and effect and to provide valuable insight into the design and performance of equipment.

- *Design of a new steering wheel for a vehicle*. The torque needed to accomplish a specific performance response is specified, and the PD formulation can be used to simulate how various humans will perform the turn. Furthermore, specific avatars can be chosen to emulate a specific targeted population in terms of anthropometry and strength.
- *Design of new levers in construction equipment*. Predictive dynamics is used to simulate force reactions at both hands and postural motion as the operator attempts to exert a certain force on the joystick control.
- *Simulation of stability while a person is wearing loads*. Predictive dynamics can be used to test the design of equipment that may restrict a person's range of motion or may add a balanced or unbalanced load, changing their stability. In PD, the notion of stability was effectively assessed using the ZMP approach and was then dynamically adapted.

10.2.6 Egress/ingress

Entering and exiting a vehicle, a building, or any other constrained space has been a challenge to analyze for many years. Consider, for example, a new vehicle that has its interior ergonomically designed for maximum comfort, but may be difficult to ingress. The door opens, but the constrained access requires the occupant to assume quite uncomfortable postures, particularly in a dynamic manner, where the body weight is sustained by the arms holding onto the vehicle's structure. This swing motion needed to enter the vehicle will result in significant stress on some joints. This is particularly true in large equipment used for farming, military operations, construction, etc.

Predictive dynamics allows for simulating these conditions and trying various body sizes, weights, and strength capabilities while considering the geometric design of the vehicle. Several simulations can be done to assess the various difficulties, change the design, and rerun the simulation until a set of solutions is deemed acceptable.

10.2.7 Unsafe situations

In many situations, humans are sent into harm's way to test a prototype. PD can be used to mitigate the risk of testing unsafe designs or situations. Consider an amusement park ride that requires occupants of various anthropometries, weights, ages, and strengths to be subjected to strenuous dynamics. The prototypes of these rides typically take several iterations with complete fabrication, using significant time and resources. They are typically tested with dummies representing humans until the ride is deemed safe for a first human attempt. There are many other unsafe conditions that require human action and are characterized by human dynamics. In these cases, it is always safer, more effective, and less costly to send the virtual human into the virtual world. Predictive dynamics can be used to significantly reduce the costs of design, prototyping, and time to market.

10.3 Future research

The authors believe that the field of digital human modeling is only at its beginning stages of maturity. While we have presented a rigorous method for predicting human motion, we believe that many disciplines are needed to make up a digital human model. As each of these disciplines matures, it will significantly contribute to this science.

10.3.1 Soft-tissue dynamics

Predictive dynamics has been introduced in this book as a method that uses rigid-body dynamics to simulate the motion of the human body. It is understood that the body has significant soft tissue that allows for a gentler and softer impact with the ground, for example. Modeling soft tissue is the subject of many research efforts, whether it is muscle flexibility, wrapping, and sliding, or skin deformation under motion that affects the range of motion. Predictive dynamics does allow for the expansion of the theory to include soft-tissue deformation.

Meshless methods for finite element analysis and non-uniform rational B-splines (NURBS) superimposed by constitutive equations have led to significant recent advances in addressing soft-tissue mechanics. Our team (Lu, 2011), for example, has made significant advances in the study of clothing deformation, which is characterized by highly nonlinear behavior. These investigations, we believe, will lead to a far better understanding of soft-tissue mechanics and interactions with the human body.

10.3.2 Intelligence

The PD model uses cost functions to drive the motion and, as a result, induces behavior. This behavior has been shown to vary depending on the task, the situation, and the physical and environmental conditions. One goal of future research is to better understand which functions are cognitively selected. The introduction of intelligence into the PD model would add a significant component in terms of behavior and realism.

10.3.3 Psychological and physiological factors

Execution of tasks under the PD formulation does not take into account any of the psychological or physiological factors. To illustrate the need for such modeling, consider a soldier performing a task that requires significant strength and courage. The psychological status of this soldier at the time of execution will have a significant effect on the task performance.

Moreover, while PD accounts for varying strength capabilities, the internal physiological state of a person must also be modeled. This includes the person's medical condition, energy capability such as VO2 max, cardiovascular condition, thermal condition, respiratory condition, etc.

Coupling of the psychological and physiological models with PD to create a more comprehensive human simulator will be a necessary step for future advancement of human simulation.

10.3.4 Modeling with a high level of fidelity

The modeling method presented in this book uses the DH parameterization method, recursive dynamics, and the PD approach to predict cause and effect. The equations of motion in this formulation were accurately represented, albeit some terms were also removed from the computational implementation. Here are some areas to work on in order to improve or increase the level of fidelity for the PD results.

a. Increase the number of DOF representing each joint.
b. Increase the degree of coupling between joints that are coupled.
c. Incorporate a more accurate biomechanical model at each joint. For example, rather than modeling the knee joint as one DOF with a specific ROM, it is possible to investigate a more accurate kinematic model in which ligaments are represented and the resulting motion is correspondingly specified.
d. A more detailed model of the musculoskeletal system. While the PD approach employs a joint-space representation of the skeletal system whereby muscle action is represented by a resultant torque, muscle interaction and recruitment are not modeled.

10.3.5 Real-time simulation

The goal of creating a human simulator that can respond to real and active forces and moments, that can interact with researchers, and that has biomechanics that can be monitored is close and within reach. While PD offers a broad underlying formulation for achieving this goal, there are still several obstacles. One obstacle is the inability to execute PD in real time because it is so computationally intensive. Both the equations of motion and the optimization process are numerically costly. Over the years, PD has been expedited by changing the algorithm to a faster code, by advancing the computational platforms, and by parallelizing some of the code structure. Even though simulation of some of the tasks has been reduced to a few minutes on a fast CPU, the main obstacle is in further parallelizing the optimization code, as it has to be sequentially executed. Further research in this area, more efficient solvers, less complicated formulations, and other aspects will be considered to enhance its computational efficiency.

Reference

Lu, J., 2011. Isogemetric contact analysis: geometric basis and formulation for frictionless contact. Comput. Methods in Appl. Mech. Eng. 200, 726–741.

Bibliography

Abdel-Malek, K., Yu, W., 2004. A mathematical method for ergonomic-based design. Int. J. Ind. Ergon. 34 (5), 375–394.

Abdel-Malek, K., Yu, W., Jaber, M., 2001. Realistic Posture Prediction. 2001 SAE Digital Human Modeling and Simulation.

Abdel-Malek, K., Wei, Y., Mi, Z., Tanbour, E., Jaber, M., 2001a. Posture prediction versus inverse kinematics. In: Proceedings of the 2001b. ASME Design Engineering Technical Conferences and Computers and Information in Engineering Conference, Pittsburgh, PA, pp. 37–45.

Abdel-Malek, K., Yang, J., Brand, R., Tanbour, E., 2001c. Towards understanding the workspace of the upper extremities. SAE Trans. J. Passenger Cars: Mech. Syst. 110 (6), 2198–2206.

Abdel-Malek, K., Yu, W., Jaber, M., Duncan, J., 2001d. Realistic posture prediction for maximum dexterity. SAE Technical Paper 2001-01-2110. doi: 10.427/2001-01-2110.

Abdel-Malek, K., Yu, W., Jaber, M., Duncan, J., 2002. Realistic posture prediction for maximum dexterity. SAE Trans. J. Passenger Cars: Mech. Syst. 110 (6), 2241–2249.

Abdel-Malek, K., Yang, J., Brand, R., Tanbour, E., 2004. Towards understanding the workspace of human limbs. Ergonomics 47 (13), 1386–1406.

Abdel-Malek, K., Yang, J., Yu, W., Duncan, J., 2004a. Human performance measures: mathematics. In: Proceedings of the ASME Design Engineering Technical Conferences (DAC 2004). Salt Lake City, UT.

Abdel-Malek, K., Yang, J., Mi, Z., Patel, V.C., Nebel, K., 2004b. Human upper body motion prediction. Proceedings of Conference on Applied Simulation and Modeling (ASM) 2004, Rhodes, Greece.

Abdel-Malek, K., Yang, J., Kim, J., Marler, T., Beck, S., Nebel, K., 2004c. Santos: a virtual human environment for human factors assessment. In: Paper presented at the 24th Army Science Conference: Transformational Science and Technology for the Current and Future Force, Orlando, FL, pp. 28–30.

Abdel-Malek, K., Yang, J., Marler, T., Beck, S., Mathai, A., Zhou, X., et al., 2006. Towards a new generation of virtual humans. Int. J. Hum. Factors Model. Simul. 1 (1), 2–39.

Abdel-Malek, K., Mi, Z., Yang, J., Nebel, K., 2006. Optimization-based trajectory planning of human upper body. Robotica 24 (6), 683–696.

Abdel-Malek, K., Arora, J., Yang, J., Marler, T., Beck, S., Swan, S., et al., 2006. Santos: a physics-based digital human simulation environment. In: Paper presented at the 50th Annual Meeting of the Human Factors and Ergonomics Society, San Francisco, CA.

Abdel-Malek, K., Yang, J., Marler, T., Beck, S., Mathai, A., Zhou, X., et al., 2006. Towards a new generation of virtual humans. Int. J. Hum. Factors Model. Simul. 1 (1), 2–39.

Abdel-Malek, K., Yang, J., Kim, J.K., Marler, T., Beck, S., Swan, C., et al., 2007. Development of the virtual human Santos. In: Proceedings of the First International Conference on Digital Human Modeling, Proceedings. Springer-Verlag, Berlin, pp. 490–499.

Abdel-Malek, K., Arora, J., Frey-Law, L., Swan, C., Beck, S., Xia, T., et al., 2008. Santos: a digital human in the making. In: Proceedings of the IASTED International Conference on Applied Simulation and Modeling, Corfu, Greece, June.

Anderson, F.C., Pandy, M.G., 2001a. Static and dynamic optimization solutions for gait are practically equivalent. J. Biomech. 34 (2), 153–161.

Anderson, F.C., Pandy, M.G., 2001b. Dynamic optimization of human walking. J. Biomech. Eng. 123 (5), 381–390.

Andriacchi, T.P., 2002. Increased knee joint loads during walking are present in subjects with knee osteoarthritis. Osteoarthritis Cartilage 10, 573–579.

Arora, J.S., 1990. Computational design optimization: a review and future directions. Struct. Saf. 7 (2), 131–148.

Arora, J.S., 1997. Guide to Structural Optimization. American Society of Civil Engineers, Reston, VA.

Arora, J.S., 1999. Optimization of structures subjected to dynamic loads. In: Leondes, C. T. (Ed.), Structural Dynamic Systems Computational Techniques and Optimization, vol. 7. Gordon and Breach Science Publishers, pp. 1–73.

Arora, J.S., 2002. Methods for discrete variable structural optimization. In: Burns, S. (Ed.), Recent Advances in Optimal Structural Design. Structural Engineering Institute, ASCE, Reston, VA, pp. 1–40.

Arora, J.S., 2007. Optimization of Structural and Mechanical Systems. World Scientific Publishing Co., Hackensack, NJ.

Arora, J.S., Baenziger, G., 1986. Uses of artificial intelligence in design optimization. Comput. Methods Appl. Mech. Eng. 54 (3), 303–323.

Arora, J.S., Tseng, C.H., 1988. Interactive design optimization. Eng. Optimiz. 13 (3), 173–188.

Aspelin, K., 2005. Establishing Pedestrian Walking Speeds. Portland State University. Retrieved 2009-08-24.

Avin, K., Frey-Law, L.A., 2009. Endurance time is joint-specific: a modeling and meta-analysis investigation. In: Paper presented at the American Society for Biomechanics Annual Scientific Meeting, Penn State University, PA.

Avin, K., Gentile, A.J., Ford, B., Moore, H., Monitto-Webber, M., Naughton, M., et al., 2009. Sex differences in fatigue at the elbow and ankle. In: Paper presented at the APTA Section on Research Retreat, CA.

Avin, K., Naughton, M.R., Ford, B.W., Moore, H.E., Monitto-Webber, M.N., Stark, A.M., et al., 2010. Sex differences in fatigue resistance are muscle group dependent. Med. Sci. Sports Exerc. 42 (10), 1943–1950.

Bataineh, M.H., 2012. Artificial Neural Network for Studying Human Performance (MS Thesis). The University of Iowa.

Beck, D.J., Chaffin, D.B., 1992. An evaluation of inverse kinematics models for posture prediction. Computer Applications in Ergonomics, Occupational Safety and Health. pp. 329–336.

Bhatt, R., Xiang, Y., Kim, J., Mathai, A., Penmatsa, R., Chung, H-J., et al., 2008. Dynamic optimization of human stair-climbing motion. In: Paper presented at the SAE Digital Human Modeling Conference, Pittsburgh, PA.

Bhatti, M., Vignes, R., Han, R., Horn, N., 2004. Incorporating muscle fatigue in a virtual soldier environment. (Abstract no. 1546). In: Proceedings of the International Soldier Systems Conference, Boston. MA.

Bhatti, M., Vignes, R., Han, R., 2005. Muscle Forces and Fatigue in a Digital Human Environment (SAE Paper No. 05DHM-73), SAE International, Warrendale, PA.

Biewener, A.A., 2003. Animal Locomotion. Oxford University Press, USA.

Bottasso, C.L., Prilutsky, B.I., Croce, A., Imberti, E., Sartirana, S., 2006. A numerical procedure for inferring from experimental data the optimization cost functions using a multi-body model of the neuro-musculoskeletal system. Multibody Syst. Dyn. 16, 123–154.

Cappozzo, A., Catani, F., Della Croce, U., Leardini, A., 1995. Position and orientation in space of bones during movement: anatomical frame definition and determination. Clin. Biomech. 10 (4), 171–178.

Carrolll, D.L., web page <http://cuaerospace.com/carroll/ga.html/>. (accessed 25-04-2012).

Chang, C.C., Brown, D.R., Bloswick, D.S., Hsiang, S.M., 2001. Biomechanical simulation of manual lifting using space-time optimization. J. Biomech. 34, 527–532.

Cheng, H., Obergefell, L., Rizer, A., 1994. Generator of Body (GEBOD) Manual, AL/CF-TR-1994-0051, Armstrong Laboratory, Wright-Patterson Air Force Base, Ohio.

Chevallereau, C., Formal'sky, A., Perrin, B., 1998. Low energy cost reference trajectories for a biped robot. In: Proceedings of IEEE International Conference on Robotics and Automation, vol. 2. Leuven, Belgium, pp. 1398–1404.

Chiaverini, S., Siciliano, B., 1991. Redundancy resolution for the human-arm-like manipulator. Rob. Auton. Syst. 8, 239–250.

Chung, H-J., Xiang, Y., Mathai, A., Rahmatalla, S., Kim, J., Marler, T., et al., 2007. A Robust Formulation for Prediction of Human Running (SAE Paper No. 2007-01-2490). SAE International, Warrendale, PA.

Coeazz, S., Mundermann, L., Andriacchi, T., 2007. A framework for functional identification of joint centers using markerlaess motion capture, validation for the hip joint. J. Biomech. 40, 3510–3515.

Das, B., Behara, D.N., 1998. Three-dimensional workspace for industrial workstations. Hum. Factors 40 (4), 633–646.

Das, B., Sengupta, A.K., 1995. Computer-aided human modeling programs for workstation design. Ergonomics 38, 1958–1972.

Della Croce, U., Cappozzo, A., Kerrigan, D.C., 1999. Pelvis and lower limb anatomical landmark calibration precision and its propagation to bone geometry and joint angles. Med. Biol. Eng. Comput. 37, 155–161.

Faraway, J.J., 1997. Regression analysis for a functional response. Techometrics 39 (3), 254–262.

Farrell, K., Marler, R.T., 2004. Optimization-Based Kinematic Models for Human Posture. The University of Iowa, Virtual Soldier research Program, Technical Report Number VSR-04.11.

Farrell, K., Yang, J., Abdel-Malek, K., 2004. Santos: a new interactive virtual human. Paper presented at SIGGRAPH 2004 Real-Time 3DX: Demo or Die, Los Angeles, CA.

Farrell, K., Marler, R.T., Abdel-Malek, K., 2005. Modeling dual-arm coordination for posture: an optimization-based approach. SAE Human Modeling for Design and Engineering Conference, Iowa City, IA.

Fregly, B.J., Reinbolt, J.A., Rooney, K.L., Mitchell, K.H., Chmielewski, T.L., 2007. Design of patient-specific gait modifications for knee osteoarthritis rehabilitation. IEEE T. Bio.-Med. Eng. 54 (9), 1687–1695.

Frey-Law, L.A., Laake, A., Delmonaco, C., 2007. Contributions of passive-tension vs. inertial effects on gravity correction for strength training. In: Paper presented at the American Society for Biomechanics, Palo Alto, CA.

Frey-Law, L.A., Lee, J., McMullen, T., Baier, T., Goodall, K., McEchron, C., 2008. Perceived pain and exertion during fatigue are distinct. In: Paper presented at the APTA Combined Sections Meeting, Nashville, TN.

Frey-Law, L.A., Krishnan, C., Avin, K.G., 2010. Modeling nonlinear errors in surface electromyography due to baseline noise: A new methodology. J. Biomech. 44 (1), 202−205.

Frey-Law, L.A., Lee, J.E., McMullen, T., Xia, T., 2010. Relationships between maximum holding time and ratings of pain and exertion differ for static and dynamic tasks. Appl. Ergon. 42 (1), 9−15.

Fujimoto, H., Zhu, J., Abdel-Malek, K., 2001. Image-based visual servoing for optimal grasping. J. Rob. Mechatronics 13 (5), 279−487.

Goldenberg, A.A., Benhabib, B., Fenton, R.G., 1985. A complete generalized solution to the inverse kinematics of robots. IEEE J. Rob. Autom. 1 (1), 14−20.

Goussous, F., Marler, T., Abdel-Malek, K., 2009. A new methodology for human grasp prediction. IEEE T. Syst., Man, Cybern., Part A: Syst. Hum. 39 (2), 369−380.

Gundogdu, O., Anderson, K.S., Parnianpour, M., 2005. Simulation of manual materials handling: Biomechanial assessment under different lifting conditions. Technol. Health Care. 13 (1), 57–66.

Hagio, K., Sugano, N., Nishii, T., Miki, H., Otake, Y., Hattori, A., et al., 2004. A novel system of four-dimensional motion analysis after total hip athroplasty. J. Orthop. Res. 22 (3), 665−670.

Hariri, M., 2012. A Study of Optimization-Based Predictive Dynamics Method for Digital Human Modeling (PhD Dissertation), The University of Iowa.

Hariri, M., Bhatt, R., Arora, J., Abdel-Malek, K., 2010. Optimization-based collision avoidance using spheres, finite cylinders, and finite planes. In: Paper presented at the 3rd International Conference on Applied Human Factors and Ergonomics (AHFE). Miami, FL.

Hariri, M., Arora, J., Abdel-Malek., K., 2013. Optimization-based prediction of a soldier's motion: stand-prone-aim task. In: Viadero, F., Ceccarelli, M. (Eds.), I New Trends in Mechanism and Machine Science, Mechanism and Machine Science 7, Springer Science+business Media, Dordrecht, pp. 459−467.

Horn, E., 2005. Optimization Based Dynamic Human Motion Prediction (MS Thesis). The University of Iowa.

Hsiang, S.M., Chang, C.C., McGorry, R.W., 1999. Development of a set of equations describing joint trajectories during para-sagittal lifting. J. Biomech. 32, 871−876.

Hunstad, T., Lee, J., Frey-Law, L., 2007. Perceived exertion versus muscle activation strategy during isometric elbow flexion fatigue. Paper presented at the APTA Combined Sections Meeting, Boston, MA.

Inman, V.T., Ralston, R.J., Todd, F., 1981. Human Walking. Wilkins and Wilkins, Baltimore, MD.

Johnson, R., Smith, B.L., Penmatsa, R., Marler, T., Abdel-Malek, K., 2009. Real-time obstacle avoidance for posture prediction. In: Proceedings of the SAE Digital Human Modeling Conference, Goteborg. Sweden.

Johnson, R., Fruehan, C., Schikore, M., Marler, T., Abdel-Malek, K., 2010. New developments with collision avoidance for posture prediction. In: Proceedings of the Third International Conference on Applied Human Factors and Ergonomics, Miami, FL.

Jung, E.S., Kee, D., Chung, M.K., 1992. Reach posture prediction of upper limb for ergonomic workspace evaluation. In: Proceedings of the 36th Annual Meeting of the Human Factors Society. Atlanta, GA, Part, Vol. 1, pp.702–706.

Jung, E.S., Kee, D., Chung, M.K., 1995. Upper body reach posture prediction for ergonomic evaluation models. Int. J. Ind. Ergon. 16, 95–107.

Kee, D., Jung, E.S., Chang, S., 1994. A man-machine interface model for ergonomic design. Comput. Ind. Eng. 27, 365–368.

Kim, J., Abdel-Malek, K., Mi, Z., Nebel, K., 2004. Layout design using an optimization-based human energy consumption formulation. In: Paper presented at SAE Digital Human Modeling for Design and Engineering, Rochester, MI.

Kim, J., Abdel-Malek, K., Yang, J., Nebel, K., 2004. Dynamic motion prediction and energy level determination for a virtual soldier's upper body. In: Paper presented at the International Soldier Systems Conference, Boston, MA.

Kim, J., Yang, J., Abdel-Malek, K., Nebel, K., 2005. Task-based vehicle interior layout design using optimization method to enhance safety. In: Trevisani, D., Sisti, A. (Eds.), Enabling Technologies for Simulation Science IX. SPIE, Bellingham, WA, pp. 54–65.

Kim, J., Abdel-Malek, K., Yang, J., Nebel, K., 2005. Optimization-based dynamic motion simulation and energy consumption prediction for a digital human. J. Passenger Car: Electron. Electronical Syst. 114 (7), 797–806.

Kim, J., Abdel-Malek, K., Yang, J., Marler, T., Nebel, K., 2005. Lifting posture analysis in material handling using virtual humans (Paper No. IMECE2005-81801). In: Paper presented at the 2005 ASME International Mechanical Engineering Congress and Exposition, Orlando, FL.

Kim, J., Abdel-Malek, K., Yang, J., Nebel, K., 2006. Motion prediction and inverse dynamics for human upper extremities. In: Paper presented at the 2006 SAE DHM Conference, Detroit, MI.

Kim, J., Yang, J., Abdel-Malek, K., 2007. Load-effective dynamic motion planning for redundant manipulators (Paper No. DETC2007-35393). In: Paper presented at the ASME 2007 International Design Engineering Technical Conferences and Computers and Information in Engineering Conference, Las Vegas, NV.

Kim, J., Xiang, Y., Bhatt, R., Yang, J., Chung, H.J., Patrick, A., et al., 2008. General biped motion and balance of a human model (SAE Paper No. 2008-01-1932). SAE International, Warrendale, PA.

Kim, J., Yang, J., Abdel-Malek, K., 2008. A novel formulation for determining joint constraint loads during optimal dynamic motion of redundant manipulators in DH representation. Multibody Syst. Dyn. 19, 427–451.

Kim, J.G., Baek, J.H., Park, F.C., 1999. Newton-type algorithms for robot motion optimization. Proc. IEEE/RSJ Int. Conf. Intell. Robots Syst. 3, 1842–1847.

Kim, J.H., Abdel-Malek, K., Yang, J., Marler, R.T., 2006. Prediction and analysis of human motion dynamics performing various tasks. International Journal of Human Factors Modelling and Simulation 1 (1), 69–117.

Kim, J.H., Abdel-Malek, K., Xiang, Y., Yang, J., Arora, J.S., 2011. Concurrent motion planning and reaction load distribution for redundant dynamic systems under external holonomic constraints. Int. J. Numer. Methods Eng. 87, n/a. doi: 10.1002/nme.3162.

Knake, L., Mathai, A., Marler, T., Farrell, K., Johnson, R., Abdel-Malek, K., 2010. New capabilities for vision-based posture prediction. In: Paper presented at the 3rd International Conference on Applied Human Factors and Ergonomics, Miami, FL.

Kramer, J.F., Balsor, B.E., 1990. Lower-extremity preference and knee extensor torques in intercollegiate soccer players. Can. J. Sport Sci. [*Revue Canadienne Des Sciences Du Sport*] 15 (3), 180−184.

Krishnakumar, K., 1989. Micro-genetic algorithms for stationary and non-stationary function optimization. SPIE Intelligent Control and Adaptive Systems, 1196, pp. 289−296.

Laake, A., Frey-Law, L.A., 2007. Modeling 3D knee torque surfaces for males and females. In: Paper Presented at the American Society for Biomechanics, August, Palo Alto, CA.

Leboeuf, F., Bessonnet, G., Seguin, P., Lacouture, P., 2006. Energetic versus sthenic optimality criteria for gymnastic movement synthesis. Multibody Syst. Dyn. 16 (3), 213−236.

Lee, S.H., Kim, J., Park, F.C., Kim, M., Bobrow, J.E., 2005. Newton-type algorithm for dynamics-based robot movement optimization. IEEE Trans. Rob. 21 (4), 657−667.

Limerick, R.B., Abernethy, B., 1997. Qualitatively different modes of manual lifting. Int. J. Ind. Ergon. 19, 413−417.

Lin, C.J., Ayoub, M.M., Bernard, T.M., 1999. Computer motion simulation for sagittal plane lifting activities. Int. J. Ind. Ergon. 24, 141−155.

Liu, Q., Marler, T., Yang, J., Kim, J., Harrison, C., 2009. Posture prediction with external loads—a pilot study. SAE Int. J. Passenger Cars: Mech. Syst. 2 (1), 1014−1023.

Lu, J., Zhao, X., 2009. Pointwise identification of elastic properties in nonlinear hyperelastic membranes. Part I: theoretical and computational developments. ASME J. Appl. Mech. 76, Paper no. 061013, 10 pages. doi 10.1115/1.3130805.

Man, X., Swan, C., 2004. Mathematical clothing modeling in a digital human environment. In: Paper Presented at 2004 IMECE International Mechanical Engineering Congress and RandD Exposition, Anaheim, CA.

Marler, R.T., Arora, J.S., 2004a. Survey of multi-objective optimization methods for engineering. Struct. Multi. Optim. 26 (6), 369−395.

Marler, R.T., Arora, J.S., 2004b. Study of multi-objective optimization using simplified crash models. In: 10th AIAA/ISSMO Multidisciplinary Analysis and Optimization Conference, August, Albany.

Marler, R.T., Arora, J.S., 2010. The weighted sum method for multi-objective optimization: some insights. Struct. Multi. Optim. 41 (6), 853−862.

Marler, R.T., Rahmatalla, S., Shanahan, M., Abdel-Malek, K., 2005. A new discomfort function for optimization-based posture prediction. In: Paper presented at the SAE Human Modeling for Design and Engineering Conference, Iowa City, IA.

Marler, R.T., Yang, J., Arora, J.S., Abdel-Malek, K., 2005. Study of bi-criterion upper body posture prediction using pareto optimal sets. Paper presented at the IASTED International Conference on Modeling, Simulation, and Optimization, Oranjestad, Aruba.

Marler, R.T., Farrell, K., Kim, J., Rahmatalla, S., Abdel-Malek, K., 2006. Vision performance measures for optimization-based posture prediction. In: Paper presented at the SAE Human Modeling for Design and Engineering Conference, Lyon, France.

Marler, R.T., Yang, J., Rahmatalla, S., Abdel-Malek, K., Harrison, C., 2007. Validation methodology development for predicted posture. Paper Presented at the SAE Digital Human Modeling Conference, Seattle, WA.

Marler, R.T., Arora, J., Beck, S., Lu, J., Mathai, A., Patrick, A., et al., 2009. Computational approaches in digital human modeling. In: Duffy, V.G. (Ed.), Handbook of Digital Human Modeling for Human Factors and Ergonomics. CRC Press, Boca Raton, FL (Chapter 9, 17 pages).

Marler, R T., Knake, L., Johnson, R., 2011. Optimization-based posture prediction for analysis of box-lifting tasks. Paper Presented at the 3rd International Conference on Digital Human Modeling. Orlando, FL.

Mathai, A., Marler, R.T., Farrell, K., Meusch, J., Taylor, A., Beck, S., et al., 2010. A new armor simulation and evaluation toolkit. Paper Presented at the Personal Armor Systems Symposium, Quebec City, Canada.

Matsunanga, N., Okimura, Y., Kawaji, S., 2004. Kinematic analysis of human lifting movement for biped robot control. AMC '04. The 8th IEEE International Workshop on Advanced Motion Control, pp. 421−426.

McKerrow, P.J., 1991. Introduction to Robotics. Addison-Wesley Publishing Company, Sydney.

Mi, Z., Yang, J., Abdel-Malek, K., Jay, L., 2002. Planning for kinematically smooth manipulator trajectories. In: Proceedings of the 2002 ASME Design Engineering Technical Conferences and Computer and Information in Engineering Conference. American Society of Mechanical Engineers, New York, pp. 1065−1073.

Mi, Z., Yang, J., Abdel-Malek, K., Mun, J.H., Nebel, K., 2002. Real-time inverse kinematics for humans. In: Proceedings of the 2002 ASME Design Engineering Technical Conferences and Computer and Information in Engineering Conference, 5A, September, Montreal, Canada, American Society of Mechanical Engineers, New York, 349−59.

Miller, C., Mulavara, A., Bloomberg, J., 2002. A quasi-static method for determining the characteristic of motion capture camera system in a "split-volume" configuration. Gait Posture 16 (3), 283−287.

Nakamura, Y., 1991. Advanced Robotics Redundancy and Optimization. Addison-Wesley Publishing Company, Sydney.

Neumann, W.P., Wells, R.P., Norman, R.W., Kerr, M.S., Frank, J., Shannon, H.S., 2001. Trunk posture: reliability, accuracy, and risk estimates for low back pain from a video-based assessment method. J. Ind. Ergon. 28, 355−365.

Nicolas, G., Multon, F., Berillon, G., Marchal, F., 2007. From bone to plausible bipedal locomotion using inverse kinematics. J. Biomech. 40, 1048−1057.

O'Connor, T.W., 1999. Bone position estimation from skin marker coordinates using global optimization with joint constraints. J. Biomech. 32, 129−134.

Pierce, G. and Frey-Law, L.A., 2008. 3-D strength surfaces for shoulder internal and external rotation. In: Paper presented at the North American Congress on Biomechanics (NACOB), Ann Arbor, MI.

Pierce, G., Frey-Law L.A., 2008. Isokinetic 3D shoulder strength assessment: internal external rotation strength surfaces. In: Paper Presented at the Biomedical Engineering Society Conference. St. Louis, MO.

Pitarch, E.P., Yang, J., Abdel-Malek, K., Marler, R.T., 2005. Hand grasping strategy for virtual humans. Paper Presented at the 3rd IASTED International Conference on Biomechanics, Benidorm, Spain.

Rahmatalla, S., Kim, H.J., Shanahan, M., Swan, C.C., 2005. Effect of restrictive clothing on balance and gait using motion capture and stability analysis (SAE Paper No. 05DHM-45). SAE International, Warrendale, PA.

Rahmatalla, S., Xia, T., Ankrum, J., Wilder, D., Frey-Law, L., Abdel-Malek, K., et al., 2007. A framework to study human response to whole-body vibration (SAE Paper No. 2007-01-2474). SAE International, Warrendale, PA.

Rahmatalla, S., Xia, T., Contratto, M., Kopp, G., Wilder, D., Frey-Law, L., 2008. Three-dimensional motion capture protocol for seated operator in whole-body vibration. Int. J. Indust. Ergon. 38, 425–433.

Rahmatalla, S., Xiang, Y., Smith, R., Li, J., Meusch, J., Bhatt, R., et al., 2008. A validation protocol for predictive human locomotion (SAE Paper No. 08DHM-0024/2008-01-1855). SAE International, Warrendale, PA.

Rahmatalla, S., Xiang, Y., Smith, R., Meusch, J., Li, J., Bhatt, R., et al., 2009. Validation of Santos biomechanics (ASME Paper No. SBC209-204916). In: Paper Presented at the ASME 2009 Summer Bioengineering Conference. Lake Tahoe, CA.

Rahmatalla, S., Xiang, Y., Smith, R., Meusch, J., Li, J., Marler, R.T., et al., 2009. Validation of lower-body posture prediction for the virtual human model Santos (SAE Paper No. 09DHM-0027). Paper Presented at the SAE Digital Human Modeling Conference, Goteborg, Sweden.

Rahmatalla, S., Smith, R., Xia, T., Contratto, M., 2009. Discomfort measure in multiple-axis whole-body vibration. Extended abstract presented at the 4th International Conference on Whole Body Vibration Injuries, Montreal, Canada.

Rahmatalla, S., Smith, R., Meusch, J., Xia, T., Marler, R.T., Contratto, M., 2010. A quasi-static discomfort measure in whole-body vibration. Indust. Health 48 (5), 645–653.

Rahmatalla, S., Xiang, Y., Smith, R., Meusch, J., Bhatt, R., 2011. A validation framework for predictive human models. Int. J. Human Fact. Model. Simul. 2(1), 67–84.

Riffard, V., Chedmail, P., 1996. Optimal posture of a human operator and CAD in robotics. In: Proceedings of the 1996 IEEE International Conference on Robotics and Automation. April, Minneapolis, MN, Institute of Electrical and Electronics Engineers, New York, pp. 1199–1204.

Rochambeau, B., Marler, R.T., Abdel-Malek, K., 2008. Multiple user-defined end-effectors with shared memory communication for posture prediction. Paper Presented at the SAE Digital Human Modeling Conference, Pittsburgh, PA.

Roussel, L., Canudas-de-Wit, C., Goswami, A., 1998. Generation of energy optimal complete gait cycles for biped robots. In: Proceedings of IEEE International Conference on Robotics and Automation, Leuven, Belgium, 3, pp. 2036–2041.

Schiehlen, W., 1997. Multibody system dynamics: roots and perspectives. Multibody Syst. Dyn. 1, 149–188.

Sinokrot, T., Yang, J., Fetter, B., Abdel-Malek, K., 2005. Workspace analysis and visualization for Santos's upper extremity. J. Passenger Car: Mech. Syst. 114 (6), 2970–2982.

Smith, B., Marler, T., Abdel-Malek, K., 2008. Studying visibility as an objective for posture prediction. In: Paper presented at the SAE Digital Human Modeling Conference. Pittsburgh, PA.

Sohl, G.A., Bobrow, J.E., 2001. A recursive multibody dynamics and sensitivity algorithm for branched kinematic chains. ASME J. Dyn. Syst., Meas., Control 123, 391–399.

Toogood, R.W., 1989. Efficient robot inverse and direct dynamics algorithms using microcomputer based symbolic generation. In: IEEE International Conference on Robotics and Automation, pp. 1827–1832.

Uicker, J.J., 1965. On the Dynamic Analysis of Spatial Linkages Using 4×4 Matrices. Northwestern University, Evanston.

Vukobratović, M., Potkonjak, V., Babković, K., Borovac, B., 2007. Simulation model of general human and humanoid motion. Multibody Syst. Dyn. 17, 71–96.

Wang, Q., Arora, J., 2004. Alternate formulations for structural optimization. In: Paper presented at the 45th AIAA/ASME/AHS/ASC Structures, Structural Dynamics and Materials Conference. Palm Springs, CA.

Wang, Q., Arora, J., 2006. An evaluation of some alternative formulations for transient dynamic response optimization (Paper No. AIAA 2006−2052). In: Paper presented at the 47th AIAA/ASME/AHS/ASC Structures, Structural Dynamics, and Materials Conference. Newport, RI.

Wang, X.G., 1999. A behavior-based inverse kinematics algorithm to predict arm prehension postures for computer-aided ergonomic evaluation. J. Biomech. 32, 453−460.

Wang, X.G., Verriest, J.P., 1998. A geometric algorithm to predict the arm reach posture for computer-aided ergonomic evaluation. J. Vis. Comput. Animation 9, 33−47.

Wu, X., Ma, L., Chen, Z., Gao, Y., 2004. 12-DOF analytic inverse kinematics solver for human motion control. J. Inf. Comput. Sci. 1 (1), 137−141.

Xia, T., Frey-Law, L.A., 2008. Multiscale approach to muscle fatigue modeling. In: Paper presented at the Pacific Symposium on Biocomputation. Big Island, HI.

Xia, T., Frey-Law, L.A., 2008. Modeling muscle fatigue for multiple joints. In: Paper Presented at the North American Congress on Biomechanics (NACOB). Ann Arbor, MI.

Xia, T., Frey-Law, L.A., 2008. A theoretical approach for modeling peripheral muscle fatigue and recovery. J. Biomech. 41 (14), 3046−3052.

Xiang, Y., Arora, J.S, Rahmatalla, S., Abdel-Malek, K., 2009. Optimization-based dynamic human walking prediction: one step formulation. Int. J. Numer. Methods Eng. 79 (6), 667−695.

Xiang, Y., Arora, J.S., Rahmatalla, S., Bhatt, R., Marler, T., Abdel-Malek, K., 2009. Human lifting simulation using multi-objective optimization approach. Multibody Syst. Dyn. 23 (4), 431−451.

Xiang, Y., Arora, J.S., Abdel-Malek, K., 2010. Optimization-based prediction of asymmetric human gait. J. Biomech. 44 (4), 683−693.

Xiang, Y., Arora, J.S., Abdel-Malek, K., 2010. Physics-based modeling and simulation of human walking: a review of optimization-based and other approaches. Struct. Multi. Optim. 42 (1), 1−23.

Xiang, Y., Chung, H.J., Kim, J.H., Bhatt, R., Rahmatalla, S., Yang, J., et al., 2010. Predictive dynamics: an optimization-based novel approach for human motion simulation. Struct. Multi. Optim. 41 (3), 465−479.

Xiang, Y., Rahmatalla, S., Arora, J.S., Abdel-Malek, K., 2011. Enhanced optimization-based inverse kinematics methodology considering skeletal discomfort. Int. J. Hum. Factors Model. Simul. 2 (1−2), 111−126.

Xiang, Y., Arora, J.S., Abdel-Malek, K., 2012. Hybrid predictive dynamics: a new approach to simulate human motion. Multibody Syst. Dyn 28 (3), 199−224.

Xiang, Y., Arora, J., Abdel-Malek, K., 2012. 3D human lifting motion prediction with different performance measures. Int. J. Humanoid Robotics, 9 (2), Paper No. 1250012.

Xiang, Y., Arora, J., Chung, H.J., Kwon, H., Rahmatallah, S., Bhatt, R., et al., 2012. Predictive simulation of human walking transitions using an optimization formulation. Struct. Multi. Optim. 45 (5), 759−772.

Yang, J., Abdel-Malek, K., Nebel, K., 2004. Restrained and unrestrained driver reach barriers. SAE Trans.: J. Aerosp. 113 (1), 288−296.

Yang, J., Abdel-Malek, K., Farrell, K., Nebel, K., 2004. The IOWA interactive digital-human virtual environment. In: Paper Presented at the 3rd Symposium on Virtual Manufacturing and Application, Anaheim, CA.

Yang, J., Marler, R.T., Kim, H., Arora, J.S., Abdel-Malek, K., 2004. Multi-objective optimization for upper body posture prediction. In: Paper Presented at the 10th AIAA/ISSMO Multidisciplinary Analysis and Optimization Conference, Albany, NY.

Yang, J., Marler, R., Kim, H., Arora, J., Abdel-Malek, K., 2004. Multi-objective Optimization for Upper Body Posture Prediction. 10th AIAA/ISSMO Multidisciplinary Analysis and Optimization Conference, August, Albany.

Yang, J., Abdel-Malek, K., Nebel, K., 2005. On the determination of driver reach and barriers. Int. J. Vehicle Des. 37 (4), 253–273.

Yang, J., Abdel-Malek, K., Nebel, K., 2005. Reach envelope of a 9 degree of freedom model of the upper extremity. Int. J. Rob. Autom. 20 (4), 240–259.

Yang, J., Marler, R.T., Kim, H.J., Farrell, K., Mathai, A., Beck, S., et al., 2005. Santos: A new generation of virtual humans. In: Paper Presented at the SAE 2005 World Congress. Detroit, MI.

Yang, J., Pena-Pitarch, E., Kim, J., Abdel-Malek, K., 2006. Posture prediction and force/torque analysis for human hand (SAE Paper No. 2006-01-2326). In Proceedings of the SAE 2006 Digital Human Modeling for Design and Engineering Conference. Lyon, France.

Yang, J., Sinokrot, T., Abdel-Malek, K., Nebel, K., 2006. Optimization-based workspace zone differentiation and visualization for Santos (SAE Paper No. 2006-01-0696). In Proceedings of the SAE 2006 Digital Human Modeling for Design and Engineering Conference. Lyon, France.

Yang, J., Marler, T., Beck, S., Abdel-Malek, K., Kim, H.-J., 2006. Real-time optimal-reach posture prediction in a new interactive virtual environment. J. Comput. Sci. Technol. 21 (2), 189–198.

Yang, J., Marler, T., Beck, S., Kim, J., Wand, Q., Zhou, X., et al., 2006. New capabilities for the virtual human Santos. Paper presented at the SAE 2006 World Congress, Detroit, MI.

Yang, J., Kim, J.H., Abdel-Malek, K., Marler, T., Beck, S., Kopp, G.R., 2007. A new digital human environment and assessment of vehicle interior design. *Comput.-Aided Design* 39, 548–558.

Yang, J., Man, X., Xiang, Y., Kim, H., Patrick, A., Swan, C., et al., 2007. Newly Developed Functionalities for the Virtual Human Santos (SAE Paper No. 2007-01-0465). SAE International, Warrendale, PA.

Yang, J., Rahmatalla, S., Marler, T., Abdel-Malek, K., Harrison, C., 2007. Validation of predicted posture for the virtual human Santos. In: Paper presented at the 12th International Conference on Human-Computer Interaction. Beijing, China.

Yang, J., Verma, U., Penmatsa, R., Marler, T., Beck, S., Rahmatalla, S., et al., 2008. Development of a zone differentiation tool for visualization of postural comfort. In: Paper presented at the SAE 2008 World Congress. Detroit, MI.

Yang, J., Marler, T., Rahmatalla, S., 2011. Multi-objective optimization method for kinematic posture prediction: development and validation. Robotica 29, 245–253.

Yang, Q., Han, R.P.S., Frey-Law, L.A., 2006. Simulating motor units for fatigue in arm muscles in digital humans. In: Paper Presented at the 2006 Digital Human Modeling for Design and Engineering Conference. SAE, Lyon, France.

Zhang, X., Buhr, T., 2002. Are back and leg muscle strengths determinants of lifting motion strategy? Insight from studying the effects of simulated leg muscle weakness. Int. J. Indust. Ergon. 29, 161–169.

Zhang, X., Chaffin, D.B., 1996. Task effects on three-dimensional dynamic postures during seated reaching movements: an analysis method and illustration. In: Proceedings of the 1996 40th Annual Meeting of the Human Factors and Ergonomics Society, Philadelphia, PA, Part 1, Vol 1. pp. 594–598.

Zhou, X., Lu, J., 2004. Deformable solid modeling using NURBS-based finite element method. In: Paper Presented at the Iowa Academy of Science. Marshalltown, IA.

Zhou, X., Lu, J., 2005. NURBS-based Galerkin method and application to skeletal muscle modeling. In: Proceedings of the 2005 ACM Symposium on Solid and Physical Modeling, pp. 71–78.

Zhou, X., Lu, J., 2005. Biomechanical Analysis of Skeletal Muscle in an Interactive Digital Human System (SAE Paper No. 05DHM-49). SAE International, Warrendale, PA.

Chapter 1

Abdel-Malek, K., Yu, W., Jaber, M., 2001a. Realistic Posture Prediction. 2001 SAE Digital Human Modeling and Simulation.

Abdel-Malek, K., Wei, Y., Mi, Z., Tanbour, E., Jaber, M., 2001b. Posture prediction versus inverse kinematics. In: Proceedings of the 2001 ASME Design Engineering Technical Conferences and Computers and Information in Engineering Conference, Pittsburgh, PA, pp. 37–45.

Abdel-Malek, K., Yang, J., Brand, R., Tanbour, E., 2001c. Towards understanding the workspace of the upper extremities. SAE Trans. J. Passenger Cars: Mech. Syst. 110 (6), 2198–2206.

Abdel-Malek, K., Yu, W., Jaber, M., Duncan, J., 2001d. Realistic posture prediction for maximum dexterity. SAE Technical Paper 2001-01-2110. doi: 10.427/2001-01-2110.

Abdel-Malek, K., Yang, J., Brand, R., Tanbour, E., 2004a. Towards understanding the workspace of human limbs. Ergonomics 47 (13), 1386–1406.

Abdel-Malek, K., Yang, J., Yu, W., Duncan, J., 2004b. Human performance measures: mathematics. Proceedings of the ASME Design Engineering Technical Conferences (DAC 2004), Salt Lake City, UT.

Abdel-Malek, K., Yang, J., Mi, Z., Patel, V.C., Nebel, K., 2004c. Human upper body motion prediction. Proceedings of Conference on Applied Simulation and Modeling (ASM). Rhodes, Greece, pp. 28–30.

Abdel-Malek, K., Yang, J., Marler, T., Beck, S., Mathai, A., Zhou, X., Patrick, A., Arora, J.S., 2006. Towards a new generation of virtual humans. International Journal of Human Factors Modelling and Simulation 1 (1), 2–39.

Arora, J.S., 2012. Introduction to Optimum Design, third ed. Elsevier, Inc., Waltham, MA, USA.

Chapter 2

Abdel-Malek, K., Yu, W., Jaber, M., 2001a. Realistic Posture Prediction. 2001 SAE Digital Human Modeling and Simulation.

Abdel-Malek, K., Wei, Y., Mi, Z., Tanbour, E., Jaber, M., 2001b. Posture prediction versus inverse kinematics. In: Proceedings of the 2001 ASME Design Engineering Technical Conferences and Computers and Information in Engineering Conference. Pittsburgh, PA, pp. 37–45.

Abdel-Malek, K., Yang, J., Brand, R., Tanbour, E., 2001c. Towards understanding the workspace of the upper extremities. SAE Trans. J. Passenger Cars: Mech. Syst. 110 (6), 2198–2206.

Abdel-Malek, K., Yu, W., Jaber, M., Duncan, J., 2001d. Realistic Posture Prediction for Maximum Dexterity. SAE Technical Paper 2001-01-2110. doi:10.4271/2001-01-2110.

Denavit, J., Hartenberg, R.S., 1955. A kinematic notation for lower-pair mechanisms based on matrices. J. Appl. Mech. 77, 215–221.

Maurel, W., Thalmann, D., Hoffmeyer, P., Beylot, P., Gingins, P., Kalra, P., et al., 1996. A biomechanical musculoskeletal model of human upper limb for dynamic simulation. In: Computer Animation and Simulation. Springer Vienna, pp. 121–136.

Pieper, D.L., 1968. The Kinematics of Manipulators Under Computer Control. Ph.D. Thesis, Stanford University.

Chapter 3

Abdel-Malek, K., Yu, W., Jaber, M., 2001a. Realistic Posture Prediction. 2001 SAE Digital Human Modeling and Simulation.

Abdel-Malek, K., Wei, Y., Mi, Z., Tanbour, E., Jaber, M., 2001b. Posture prediction versus inverse kinematics. In Proceedings of the 2001 ASME Design Engineering Technical Conferences and Computers and Information in Engineering Conference (pp. 37–45). Pittsburgh, PA.

Abdel-Malek, K., Yang, J., Brand, R., Tanbour, E., 2001c. Towards understanding the workspace of the upper extremities. SAE Trans. J. Passenger Cars: Mech. Syst. 110.6, 2198–2206.

Abdel-Malek, K., Yu, W., Jaber, M., Duncan, J., (2001d). Realistic posture prediction for maximum dexterity. SAE Technical Paper 2001-01-2110. doi: 10.427/2001-01-2110.

Abdel-Malek, K., Yang, J., Brand, R., Tanbour, E., 2004a. Towards understanding the workspace of human limbs. Ergonomics 47 (13), 1386–1406.

Abdel-Malek, K., Yang, J., Yu, W., Duncan, J., 2004b. Human performance measures: mathematics. Proceedings of the ASME Design Engineering Technical Conferences (DAC 2004), Salt Lake City, UT.

Abdel-Malek, K., Yang, J., Mi, Z., Patel, V. C., Nebel, K., 2004c. Human upper body motion prediction. Proceedings of Conference on Applied Simulation and Modeling (ASM) 2004, Rhodes, Greece, pp. 28-30.

Abdel-Malek, K., Yang, J., Marler, T., Beck, S., Mathai, A., Zhou, X., et al., 2006. Towards a new generation of virtual humans Int. J. Human Fact. Model. Simul. 1(1), 2–39.

Abdel-Malek, K., Mi, Z., Yang, J., Nebel, K., 2005. Optimization-based layout design. J. Appl. Bionics Biomech. 2 (3/4), 187−196.

Arora, J.S., 2012. Introduction to Optimum Design, third ed. Elsevier Academic Press, San Diego, CA.

Byrd, R.H., Nocedal, J., Waltz, R.A., 2006. KNITRO: an integrated package for nonlinear optimization. In: di Pillo, G., Roma, M. (Eds.), Large-Scale Optimization. Springer-Verlag, pp. 35−59.

Case, K., Porter, J.M., Booney, M.C., 1990. SAMMIE: a man and workplace modelling system. In: Karwowski, W., Genaidy, A.M., Asfour, S.S. (Eds.), Computer-Aided Ergonomics. Taylor and Francis, New York, pp. 31–56.

Chaffin, D.B., Anderson, D.B.J., 1991. Occupational Biomechanics. Wiley, New York, NY.

Conn, A.R., Gould, N.I.M., Toint, P.L., 1992. LANCELOT—a fortran package for large-scale nonlinear optimization. Springer Series in Computational Mathematics. Springer-Verlag.

Drud, A.S., 1992. CONOPT—a large-scale GRG code. ORSA J. Comput. 6, 207−216.

Flash, T., Hogan, N., 1985. The coordination of arm movements: an experimentally confirmed mathematical model. J. Neurosci. 5(7), 1688–1703.

Gill, P.E., Murray, W., Saunders, M.A., 2002. SNOPT: an SQP algorithm for large-scale constrained optimization. Siam J Optim. 12 (4), 979−1006.

Jung, E.S., Choe, J., 1996. Human reach posture prediction based on psychophysical discomfort. Int. J. Ind. Ergon. 18, 173−179.

Marler, R.T., Rahmatalla, S., Shanahan, M., Abdel-Malek, K., 2005a. A new discomfort function for optimization-based posture prediction. Paper presented at the SAE Human Modeling for Design and Engineering Conference, Iowa City, IA.

Marler, R.T., Yang, J., Arora, J.S., Abdel-Malek, K., 2005b. Study of bi-criterion upper body posture prediction using pareto optimal sets. Paper presented at the IASTED International Conference on Modeling, Simulation, and Optimization, Oranjestad, Aruba.

Marler, R.T., Yang, J., Rahmatalla, S., Abdel-Malek, K., Harrison, C., 2009. Use of multi-objective optimization for digital human posture prediction. Eng. Optim. 41 (10), 925−943.

Mi, Z., 2004. Task-Based Prediction of Upper Body Motion, Ph.D. Dissertation, University of Iowa, Iowa City, IA.

Mi, Z., Yang, J., Abdel-Malek, K., Jay, L., 2002a. Planning for kinematically smooth manipulator trajectories. In Proceedings of the 2002 ASME Design Engineering Technical Conferences and Computer and Information in Engineering Conference (pp. 1065−73). New York: American Society of Mechanical Engineers.

Mi, Z., Yang, J., Abdel-Malek, K., Mun, J.H., Nebel, K., 2002b. Real-Time Inverse Kinematics for Humans. Proceedings of the 2002 ASME Design Engineering Technical Conferences and Computer and Information in Engineering Conference, 5A, September, Montreal, Canada, American Society of Mechanical Engineers, New York, 349−59.

Mi, Z., Jingzhou, Y., Abdel-Malek, K., 2009. Optimization-based posture prediction for human upper body. Robotica 27 (4), 607.

Porter, J.M., Case, K., Bonney, M.C., 1990. Computer workspace modeling. In: Wilson, J. R., Corlett, E.N. (Eds.), Evaluation of Human Work. Taylor and Francis, London, UK, pp. 472−499.

Yang, J., Abdel-Malek, K., Nebel, K., 2004a. Restrained and unrestrained driver reach barriers. SAE Trans. J. Aerosp. 113 (1), 288−296.

Yang, J., Abdel-Malek, K., Farrell, K., Nebel, K., 2004b. The IOWA interactive digital-human virtual environment. Paper presented at the 3rd Symposium on Virtual Manufacturing and Application, Anaheim, CA.

Yang, J., Marler, R.T., Kim, H., Arora, J.S., Abdel-Malek, K., 2004c. Multi-objective optimization for upper body posture prediction. Paper presented at the 10th AIAA/ISSMO Multidisciplinary Analysis and Optimization Conference, Albany, NY.

Yang, J., Marler, R., Kim, H., Arora, J., Abdel-Malek, K., 2004d. Multi-objective Optimization for Upper Body Posture Prediction. 10th AIAA/ISSMO Multidisciplinary Analysis and Optimization Conference, August, Albany, NY, American Institute of Aeronautics and Astronautics, Washington, DC.

Yang, J., Pena-Pitarch, E., Kim, J., Abdel-Malek, K., 2006a. Posture prediction and force/torque analysis for human hand (SAE Paper No. 2006-01-2326). In Proceedings of the SAE 2006 Digital Human Modeling for Design and Engineering Conference.

Yang, J., Sinokrot, T., Abdel-Malek, K., Nebel, K., 2006b. Optimization-based workspace zone differentiation and visualization for Santos (SAE Paper No. 2006-01-0696). In Proceedings of the SAE 2006 Digital Human Modeling for Design and Engineering Conference.

Yang, J., Marler, T., Beck, S., Abdel-Malek, K., Kim, H. -J., 2006c. Real-time optimal-reach posture prediction in a new interactive virtual environment. J. Comput. Sci. Technol. 21 (2), 189–198.

Yang, J., Marler, T., Beck, S., Kim, J., Wand, Q., Zhou, X., et al., 2006d. New capabilities for the virtual human Santos. Paper presented at the SAE 2006 World Congress, Detroit, MI.

Yang, Q., Han, R.P.S., Frey Law, L.A., 2006. Simulating motor units for fatigue in arm muscles in digital humans. Paper presented at the 2006 Digital Human Modeling for Design and Engineering Conference, SAE, Lyon, France.

Yu, W., 2001. Optimal Placement of Serial Manipulators, Ph.D. Dissertation, University of Iowa, Iowa City, IA.

Zacher, I., Bubb, H., 2004. Strength based discomfort model of posture and movement. SAE Digital Human Modelling for Design and Engineering, Rochester, MI. SAE paper 2004-01-2139.

Chapter 4

Bessonnet, G., Lallemand, J. P., 1990. Optimal trajectories of robot arms minimizing constrained actuators and travelling time. In: Robotics and Automation, 1990. Proceedings, 1990 IEEE International Conference. pp. 112–117.

Dissanayake, M.W.M.G., Goh, C.J., Phan-Thien, N., 1991. Time-optimal trajectories for robot manipulators. Robotica 9 (02), 131–138.

Fu, K.S., Gonzalez, R.C., Lee, C.S., 1987. Robotics. McGraw-Hill Book.

Furukawa, T., 2002. Time-subminimal trajectory planning for discrete non-linear systems. Eng. Optimiz. 34 (3), 219–243.

Hollerbach, J.M., 1980. A recursive Lagrangian formulation of manipulator dynamics and a comparative study of dynamics formulation complexity. IEEE T. Syst. Man Cyb. 10 (11), 730–736.

Sciavicco, L., Siciliano, B., 2000. Modelling and Control of Robot Manipulators. Springer Verlag.

Wang, Q., Xiang, Y., Kim, H., Arora, J. and Abdel-Malek, K. (2005). Alternative formulations for optimization-based digital human motion prediction (SAE Paper No. 05DHM-61). Warrendale, PA: SAE International.

Wang, Q., Xiang, Y., Arora, J. S. and Abdel-Malek, K. (2007). Alternative formulations for optimization-based human gait planning. 48th AIAA/ASME/ASCE/AHS/ASC Structures, Structural Dynamics and Materials Conference, Honolulu, Hawaii, Apr. 23−26.

Xiang, Y., Arora, J.S, Rahmatalla, S., Abdel-Malek, K., 2009a. Optimization-based dynamic human walking prediction: one step formulation. Int. J. Numer. Methods Eng. 79 (6), 667−695.

Xiang, Y., Arora, J.S., Rahmatalla, S., Bhatt, R., Marler, T., Abdel-Malek, K., 2009b. Human lifting simulation using multi-objective optimization approach. Multibody Syst. Dyn. 23 (4), 431−451.

Xiang, Y., Arora, J.S., Abdel-Malek, K., 2010a. Optimization-based prediction of asymmetric human gait. J. Biomech. 44 (4), 683−693.

Xiang, Y., Arora, J.S., Abdel-Malek, K., 2010b. Physics-based modeling and simulation of human walking: a review of optimization-based and other approaches. Struct. Multidiscip. Optim. 42 (1), 1−23.

Xiang, Y., Chung, H.J., Kim, J.H., Bhatt, R., Rahmatalla, S., Yang, J., et al., 2010c. Predictive dynamics: an optimization-based novel approach for human motion simulation. Struct. Multidiscip. Optim. 41 (3), 465−479.

Chapter 5

Arora, J.S., Wang, Q., 2005. Review of formulations for structural and mechanical system optimization. Struct. Multidisciplinary Optim. 30, 251−272.

Cahalan, T.D., Johnson, M.E., Liu, S., Chao, E.Y.S., 1989. Quantitative measurements of hip strength in different age-groups. Clin. Orthop. Relat. Res. 246, 136−145.

Gill, P.E., Murray, W., Saunders, M.A., 2002. SNOPT: an SQP algorithm for large-scale constrained optimization. Siam J. Optim. 12 (4), 979−1006.

Gubina, F., Hemami, H., McGhee, R.B., 1974. On the dynamic stability of biped locomotion. Biomed. Eng. IEEE Trans. 2, 102−108.

Huang, Q., Yokoi, K., Kajita, S., Kaneko, K., Arai, H., Koyachi, N., et al., 2001. Planning walking patterns for a biped robot. IEEE Trans. Rob. Autom. 17 (3), 280−289.

Kaminski, T.W., Perrin, D.H., Gansneder, B.M., 1999. Eversion strength analysis of uninjured and functionally unstable ankles. J. Athl. Train. 34 (3), 239−245.

Kim, H.J., Horn, E., Arora, J.S., Abdel-Malek, K., 2005. An Optimization-Based Methodology to Predict Digital Human Gait Motion. SAE International, Warrendale, PA (SAE Paper No. 05DHM-54).

Kim, H.J., Wang, Q., Rahmatalla, S., Swan, C., Arora, J., Abdel-Malek, K., et al., 2008. Dynamic motion planning of 3D human locomotion using gradient-based optimization. J. Biomech. Eng. 130 (3), 031002-1−031002-14.

Kumar, S., 1996. Isolated planar trunk strengths measurement in normals 3. Results and database. Int. J. Ind. Ergon. 17 (2), 103−111.

Rasmussen, J., Damsgaard, M., Voigt, M., 2001. Muscle recruitment by the min/max criterion—a comparative numerical study. J. Biomech. 34 (3), 409−415.

Saunders, J.B.D.M., Inman, V.T., Eberhart, H.D., 1953. The major determinants in normal and pathological gait. J. Bone Joint Surg. Am. 35-A (3), 543–558.

Vukobratović, M., Borovac, B., 2004. Zero-moment point—35 years of its life. Int. J. HR 1 (1), 157–173.

Wang, Q., Xiang, Y., Kim, H., Arora, J., Abdel-Malek, K., 2005. Alternative Formulations for Optimization-Based Digital Human Motion Prediction. SAE International, Warrendale, PA (SAE Paper No. 05DHM-61).

Xiang, Y., 2008. Optimization-based dynamic human walking prediction, PhD. Dissertation, The University of Iowa, 141 pages.

Xiang, Y., Chung, H.J., Mathai, A., Rahmatalla, S., Kim, J., Marler, T., et al., 2007. Optimization-based dynamic human walking prediction. Paper presented at the SAE Digital Human Modeling Conference, Seattle, WA.

Xiang, Y., Arora, J.S., Abdel-Malek, K., 2009a. Optimization-based motion prediction of mechanical systems: sensitivity analysis. Struct. Multidisciplinary Optim. 37 (6), 595–608.

Xiang, Y., Arora, J.S., Rahmatalla, S., Abdel-Malek, K., 2009b. Optimization-based dynamic human walking prediction: one step formulation. Int. J. Numer. Methods Eng. 79 (6), 667–695.

Xiang, Y., Arora, J.S., Rahmatalla, S., Bhatt, R., Marler, T., Abdel-Malek, K., 2009c. Human lifting simulation using multi-objective optimization approach. Multibody Syst. Dyn. 23 (4), 431–451.

Xiang, Y., Arora, J.S., Abdel-Malek, K., 2010a. Optimization-based prediction of asymmetric human gait. J. Biomech. 44 (4), 683–693.

Xiang, Y., Arora, J.S., Abdel-Malek, K., 2010b. Physics-based modeling and simulation of human walking: a review of optimization-based and other approaches. Struct. Multidisciplinary Optim. 42 (1), 1–23.

Xiang, Y., Chung, H.J., Kim, J.H., Bhatt, R., Rahmatalla, S., Yang, J., et al., 2010c. Predictive dynamics: an optimization-based novel approach for human motion simulation. Struct. Multidisciplinary Optim. 41 (3), 465–479.

Chapter 6

Anderson, D.E., Madigan, M.L., Nussbaum, M.A., 2007. Maximum voluntary joint torque as a function of joint angle and angular velocity: model development and application to the lower limb. J. Biomech. 40 (14), 3105–3113.

Bigland-Ritchie, B., Woods, J.J., 1984. Changes in muscle contractile properties and neural control during human muscular fatigue. Muscle Nerve 7 (9), 691–699.

Bohannon, R.W., 1997. Reference values for extremity muscle strength obtained by hand-held dynamometry from adults aged 20 to 79 years. Arch. Phys. Med. Rehabil. 78 (1), 26–32.

Chapman, D., Newton, M.J., Nosaka, K., 2005. Eccentric torque-velocity relationship of the elbow flexors. Isokinet. Exerc. Sci. 13 (2), 139–145.

Chapman, D.W., Newton, M.J., Zainuddin, Z., Sacco, P., Nosaka, K., 2008. Work and peak torque during eccentric exercise do not predict changes in markers of muscle damage. Br. J. Sports Med. 42 (7), 585–591.

Chow, J.W., Darling, W.G., Hay, J.G., Andrews, J.G., 1999. Determining the force-length-velocity relations of the quadriceps muscles: III. A pilot study. J. Appl. Biomech. 15 (2), 200–209.

Cleak, M.J., Eston, R.G., 1992. Delayed onset muscle soreness: Mechanisms and management. J. Sports Sci. 10 (4), 325–341.

Ding, J., Wexler, A.S., Binder-Macleod, S.A., 2000. A predictive model of fatigue in human skeletal muscles. J. Appl. Physiol. 89 (4), 1322−1332.

Ding, J., Wexler, A.S., Binder-Macleod, S.A., 2002a. A predictive fatigue model−I: predicting the effect of stimulation frequency and pattern on fatigue [erratum appears in IEEE Trans Neural Syst Rehabil Eng. 2003 Mar; 11(1):86]. IEEE Trans. Neural. Syst. Rehabil. Eng. 10 (1), 48−58.

Ding, J., Wexler, A.S., Binder-Macleod, S.A., 2002b. A predictive fatigue model−II: predicting the effect of resting times on fatigue. IEEE Trans. Neural. Syst. Rehabil. Eng. 10 (1), 59−67.

Ding, J., Wexler, A.S., Binder-Macleod, S.A., 2003. Mathematical models for fatigue minimization during functional electrical stimulation. J. Electromyogr. Kinesiol. 13 (6), 575−588.

El ahrache, K., Imbeau, D., Farbos, B., 2006. Percentile values for determining maximum endurance times for static muscular work. Int. J. Ind. Ergon. 36, 99−108.

Frey-Law, L.A., Avin, K.G., 2010. Endurance time is joint-specific: a modelling and meta-analysis investigation. Ergonomics 53 (1), 109−129.

Frey-Law, L.A., Looft, J., Heitsman, J., 2012a. A three-compartment muscle fatigue model accurately predicts joint-specific maximum endurance times for sustained isometric tasks. J. Biomech. 45 (10), 1803−1808.

Frey-Law, L.A., Laake, A., Avin, K.G., Heitsman, J., Marler, T., Abdel-Malek, K., 2012b. Knee and Elbow 3D strength surfaces: peak torque-angle-velocity relationships. J. Appl. Biomech. 28 (6), 726−737.

Gordon, A.M., Huxley, A.F., Julian, F.J., 1966. Variation in isometric tension with sarcomere length in vertebrate muscle fibres. J. Physiology−London 184 (1), 170−192.

Griffin, J.W., 1987. Differences in elbow flexion torque measured concentrically, eccentrically, and isometrically. Phys. Ther. 67 (8), 1205−1208.

Hill, A.V., 1938. The heat of shortening and the dynamic constants of muscle. Proc. R. Soc. Lond. Ser. B-Biol. Sci. 126 (843), 136−195.

Horstmann, T., Maschmann, J., Mayer, F., Heitkamp, H.C., Handel, M., Dickhuth, H.H., 1999. The influence of age on isokinetic torque of the upper and lower leg musculature in sedentary men. Int. J. Sports Med. 20 (6), 362−367.

Khalaf, K.A., Parnianpour, M., 2001. A normative database of isokinetic upper-extremity joint strengths: towards the evaluation of dynamic human performance. Biomed. Eng. App. Basis. Comm. 13, 79−92.

Khalaf, K.A., Parnianpour, M., Sparto, P.J., Simon, S.R., 1997. Modeling of functional trunk muscle performance: Interfacing ergonomics and spine rehabilitation in response to the ADA. J. Rehab. Res. Develop. 34 (4), 459−469.

Khalaf, K.A., Parnianpour, M., Karakostas, T., 2000. Surface responses of maximum isokinetic ankle torque generation capability. J. Appl. Biomech. 16 (1), 52−59.

Khalaf, K.A., Parnianpour, M., Karakostas, T., 2001. Three-dimensional surface representation of knee and hip joint torque capability. Biomed. Eng. App. Basis. Comm. 13, 53−56.

Klass, M., Baudry, S., Duchateau, J., 2005. Aging does not affect voluntary activation of the ankle dorsiflexors during isometric, concentric, and eccentric contractions. J. Appl. Physiol. 99 (1), 31−38.

Kramer, J.F., Balsor, B.E., 1990. Lower-extremity preference and knee extensor torques in intercollegiate soccer players. Can. J. Sport. Sci. (Revue Canadienne Des Sciences Du Sport) 15 (3), 180−184.

Lieber, R.L., 2002. Skeletal Muscle Structure, Function, and Plasticity. Lippincott Williams and Wilkins, Baltimore.

Liu, J.Z., Brown, R.W., Yue, G.H., 2002. A dynamical model of muscle activation, fatigue, and recovery. Biophys. J. 82 (5), 2344–2359.

Monod, H., Scherrer, J., 1965. The work capacity of a synergistic muscle group. Ergonomics 8, 329–338.

Narici, M.V., Landoni, L, Minetti, A.E., 1992. Assessment of human knee extensor muscles stress from *in vivo* physiological cross-sectional area and strength measurements. Eur. J. Appl. Physiol. Occup. Physiol. 65 (5), 438–444.

O'Brien, T.D., Reeves, N.D., Baltzopoulos, V., Jones, D.A., Maganaris, C.N., 2010. *In vivo* measurements of muscle-specific tension in adults and children. Exp. Physiol. 95 (1), 202–210.

Patrick, A., 2007. Development of a 3D Model of the Human Arm for Real-Time Interaction and Muscle Activation Prediction (MS Thesis), The University of Iowa, Iowa City, IA.

Reeves, N.D., Narici, M.V., Maganaris, C.N., 2004. Effect of resistance training on skeletal muscle-specific force in elderly humans. J. Appl. Physiol. 96 (3), 885–892.

Rohmert, W., 1960. Ermittlung von erholungspausen für statische arbeit des menschen (Determination of relaxation breaks for static work of man). Int. Z. Angew. Physiol. Einschl. Arbeitsphysiol. 18, 123–164.

Rose, L., Ericson, M., Ortengren, R., 2000. Endurance time, pain and resumption in passive loading of the elbow joint. Ergonomics 43 (3), 405–420.

Stoll, T., Huber, E., Seifert, B., Michel, B.A., Stucki, G., 2000. Maximal isometric muscle strength: normative values and gender-specific relation to age. Clin. Rheumatol. 19 (2), 105–113.

Xia, T., Frey-Law, L.A., 2008a. Multiscale Approach to Muscle Fatigue Modeling. Paper presented at the Pacific Symposium on Biocomputation, Big Island, HI.

Xia, T., Frey-Law, L.A., 2008b. Modeling Muscle Fatigue for Multiple Joints. Paper presented at the North American Congress on Biomechanics (NACOB), Ann Arbor, MI.

Xia, T., Frey-Law, L.A., 2008c. A theoretical approach for modeling peripheral muscle fatigue and recovery. J. Biomech. 41 (14), 3046–3052.

Yu, J.G., Malm, C., Thornell, L.E., 2002. Eccentric contractions leading to DOMS do not cause loss of desmin nor fibre necrosis in human muscle. Histochem. Cell Biol. 118 (1), 29–34.

Zatsiorsky, V.M., 1998. Kinematics of Human Motion. Human Kinetics Publishers, Champaign, Illinois.

Chapter 7

Anderson, F.C., Pandy, M.G., 2001. Dynamic optimization of human walking. J. Biomech. Eng.—Trans. ASME 123 (5), 381–390.

Ayyappa, E., 1997. Normal human locomotion, part 1: basic concepts and terminology. J. Prosthet. Orthot. 9 (1), 10–17.

Chevallereau, C., Aoustin, Y., 2001. Optimal reference trajectories for walking and running of a biped robot. Robotica 19, 557–569.

Choi, M.G., Lee, J., Shin, S.Y., 2003. Planning biped locomotion using motion capture data and probabilistic roadmaps. ACM Trans. Graph. 22 (2), 182–203.

Dasgupta, A., Nakamura, Y., 1999. Making feasible walking motion of humanoid robots from human motion capture data. IEEE International Conference on Robotics and Automation, Tokyo, Japan, pp. 1044−1049.

Fregly, B.J., Reinbolt, J.A., Rooney, K.L., Mitchell, K.H., Chmielewski, T.L., 2007. Design of patient-specific gait modifications for knee osteoarthritis rehabilitation. IEEE T. Bio.-Med. Eng. 54 (9), 1687−1695.

Gill, P.E., Murray, W., Saunders, M.A., 2002. SNOPT: an SQP algorithm for large-scale constrained optimization. Siam. J. Optimiz. 12 (4), 979−1006.

Goswami, A., 1999. Postural stability of biped robots and the foot-rotation indicator (FRI) point. Int. J. Robot. Res. 18 (6), 523−533.

Huang, Q., Yokoi, K., Kajita, S., Kaneko, K., Arai, H., Koyachi, N., et al., 2001. Planning walking patterns for a biped robot. IEEE T. Robot. Autom. 17 (3), 280−289.

Kajita, S., Kanehiro, F., Kaneko, K., Fujiwara, K., Harada, K., Yokoi, K., et al., 2003. Biped walking pattern generation by using preview control of zero-moment point. IEEE International Conference on Robotics and Automation Taipei, Taiwan, pp. 1620−1626.

Kim, J., Abdel-Malek, K., Yang, J., Nebel, K., 2005. Optimization-based dynamic motion simulation and energy consumption prediction for a digital human. J. Passenger Car: Electron. Electronical Syst. 114 (7), 797−806.

Lo, J., Huang, G., Metaxas, D., 2002. Human motion planning based on recursive dynamics and optimal control techniques. Multibody Syst. Dyn. 8 (4), 433−458.

Mu, X., Wu, Q., 2003. Synthesis of a complete sagittal gait cycle for a five-link biped robot. Robotica 21, 581−587.

Park, J., Kim, K., 1998. Biped Robot Walking Using Gravity-Compensated Inverted Pendulum Mode and Computed Torque Control. IEEE International Conference on Robotics and Automation, pp. 3528−3533.

Pettre, J., Laumond, J.P., 2006. A motion capture-based control-space approach for walking mannequins. Comput. Animation Virtual Worlds 17 (2), 109−126.

Ren, L., Jones, R.K., Howard, D., 2007. Predictive modelling of human walking over a complete gait cycle. J. Biomech. 40 (7), 1567−1574.

Saidouni, T., Bessonnet, G., 2003. Generating globally optimized sagittal gait cycles of a biped robot. Robotica 21, 199−210.

Sardain, P., Bessonnet, G., 2004. Forces acting on a biped robot. Center of pressure-zero moment point. IEEE T. Syst. Man Cy. A 34 (5), 630−637.

Saunders, J.B.D.M., Inman, V.T., Eberhart, H.D., 1953. The major determinants in normal and pathological gait. J. Bone Joint Surg. Am. 35-A (3), 543−558.

Simpson, K.J., Jiang, P., 1999. Foot landing position during gait influences ground reaction forces. Clin. Biomech. 14 (6), 396−402.

Stansfield, B.W., Hillman, S.J., Hazlewood, M.E., Robb, J.E., 2006. Regression analysis of gait parameters with speed in normal children walking at self-selected speeds. Gait Posture 23 (3), 288−294.

Vukobratović, M., Borovac, B., 2004. Zero-moment point—35 years of its life. Inter. J. HR 1 (1), 157−173.

Xiang, Y., 2008. Optimization-based dynamic human walking prediction (PhD. Dissertation), The University of Iowa, 141 pages.

Xiang, Y., Rahmatalla, S., Bhatt, R., Kim, J., Chung, H.J., Mathai, A., et al., 2008. Optimization-based dynamic human lifting prediction. Paper No. 2008-01-1930. Proceedings of the SAE Digital Human Modeling Conference, Pittsburgh, PA.

Xiang, Y., Arora, J.S., Rahmatalla, S., Abdel-Malek, K., 2009a. Optimization-based dynamic human walking prediction: one step formulation. Inter. J. Numer. Methods Eng. 79 (6), 667–695.

Xiang, Y., Arora, J.S., Rahmatalla, S., Bhatt, R., Marler, T., Abdel-Malek, K., 2009b. Human lifting simulation using multi-objective optimization approach. Multibody Syst. Dyn. 23 (4), 431–451.

Xiang, Y., Arora, J.S., Abdel-Malek, K., 2010a. Optimization-based prediction of asymmetric human gait. J. Biomech. 44 (4), 683–693.

Xiang, Y., Arora, J.S., Abdel-Malek, K., 2010b. Physics-based modeling and simulation of human walking: a review of optimization-based and other approaches. Struct. Multi. Optimiz. 42 (1), 1–23.

Xiang, Y., Chung, H.J., Kim, J.H., Bhatt, R., Rahmatalla, S., Yang, J., et al., 2010c. Predictive dynamics: an optimization-based novel approach for human motion simulation. Struct. Multi. Optimiz. 41 (3), 465–479.

Yamaguchi, J., Soga, E., Inoue, S., Takanishi, A., 1999. Development of a bipedal humanoid robot control method of whole-body cooperative dynamic biped walking. IEEE International Conference on Robotics and Automation, pp. 368–374.

Chapter 8

Arisumi, H., Chardonnet, J.R., Kheddar, A., Yokoi, K., 2007. Dynamic lifting motion of humanoid robots. 2007 IEEE International Conference on Robotics and Automation, Roma, Italy, pp. 2661–2667.

Authier, M., Lortie, M., Gagnon, M., 1996. Manual handling techniques: Comparing novices and experts. Int. J. Ind. Ergonom. 17 (5), 419–429.

Dysart, M.J., Woldstad, J.C., 1996. Posture prediction for static sagittal-plane lifting. J. Biomech. 29 (10), 1393–1397.

Hariri, M., 2012. A Study of Optimization-Based Predictive Dynamics method for Digital Human Modeling, PhD Dissertation. The University of Iowa, Iowa City, IA.

Huang, C., Sheth, P.N., Granata, K.P., 2005. Multibody dynamics integrated with muscle models and space-time constraints for optimization of lifting movements. ASME International Design Engineering Technical Conferences. ASME, Long Beach, California, Paper No. DETC2005-85385.

Lee, S.H., Kim, J., Park, F.C., Kim, M., Bobrow, J.E., 2005. Newton-type algorithm for dynamics-based robot movement optimization. IEEE Trans. Rob. 21 (4), 657–667.

Marler, R.T., Arora, J.S., 2005. Function-transformation methods for multi-objective optimization. Eng. Optimiz. 37 (6), 551–570.

Xiang, Y., Arora, J.S, Rahmatalla, S., Abdel-Malek, K., 2009a. Optimization-based dynamic human walking prediction: one step formulation. Int. J. Numer. Methods Eng. 79 (6), 667–695.

Xiang, Y., Arora, J.S., Rahmatalla, S., Bhatt, R., Marler, T., Abdel-Malek, K., 2009b. Human lifting simulation using multi-objective optimization approach. Multibody Syst. Dyn. 23 (4), 431–451.

Xiang, Y., Arora, J., Abdel-Malek, K., 2012. 3D human lifting motion prediction with different performance measures. Int. J. Humanoid Robotics, 9 (2), 1250012.

Zhang, X., Nussbaum, M.A., Chaffin, D.B., 2000. Back lift versus leg lift: an index and visualization of dynamic lifting strategies. J. Biomech. 33 (6), 777–782.

Chapter 9

Abdoli-Eramaki, M., Stevenson, J.M., Agnew, M.J., Kamalzadeh, A., 2009. Comparison of 3D dynamic virtual model to link segment model for estimation of net L4/L5 reaction moments during lifting. Comput. Methods Biomech. Biomed. Eng. 12 (2), 227−237.

Bartlett, D., Li, K., Zhang, X., 2007. A relation between dynamic strength and manual materials handling strategy affected by knowledge of strength. Hum. Factors 49, 438−446.

Blankevoort, L., Huiskes, R., 1996. Validation of a three-dimensional model of the knee. J. Biomech. 29 (7), 955−961.

Burgess-Limerick, R., Abernethy, B., Neal, R.J., Kippers, V., 1995. Self-selected manual lifting technique: functional consequences of the interjoint coordination. Hum. Factors 3 (2), 395−411.

Chaffin, D.B., 2002. On simulating human reach motions for ergonomic analysis. Hum. Factor Ergon. Man. 12 (3), 235−247.

Chaffin, D.B., 2009. The evolving role of biomechanics in prevention of overexertion injuries. Ergonomics 52 (1), 3−14.

Dubowsky, S.R., Rasmussen, J., Sisto, S.A., Langrana, N.A., 2008. Validation of a musculoskeletal model of wheelchair propulsion and its application to minimizing shoulder joint forces. J. Biomech. 41 (14), 2981−2988.

Garling, E.H., Kaptein, B.L., Mertens, B., Barendregt, W., Veeger, H.E.J., Neilissen, R.G. H.H., et al., 2007. Soft-tissue artefact assessment during step-up using fluoroscopy and skin-mounted markers. J. Biomech. 40, S18−S24.

Karduna, A.R., McClure, P.W., Michener, L.A., Sennett, B., 2001. Dynamic measurements of three-dimensional scapular kinematics: a validation study. J. Biomech. Eng. 123 (2), 184−190.

Lucchetti, L., Cappozzo, A., Cappello, A., Della Croce, U., 1998. Skin movement artefact assessment and compensation in the estimation of knee joint kinematics. J. Biomech. 31, 977−984.

Li, K., Zhang, X., 2009. Can relative strength between the back and knees differentiate lifting strategy? Human Factors 51 (6), 785−796.

Lu, T.W., O'connor, J.J., 1999. Bone position estimation from skin marker co-ordinates using global optimisation with joint constraints. J. Biomech. 32 (2), 129.

Marras, W.S., Sommerich, C.M., 1991. A three-dimensional motion model of loads on the lumbar spine: II. Model validation. Human Factors 33 (2), 139−149.

Matsunaga, N., Okimura, Y., Kawaji, S., 2004. Kinematic analysis of human lifting movement for biped robot control, 2004. AMC '04. The 8th IEEE International Workshop on Advanced Motion Control. pp. 421−426.

Nadzadi, M.E., Pedersen, D.R., Yack, H.J., Callaghan, J.J., Brown, T.D., 2003. Kinematics, kinetics, and finite element analysis of commonplace maneuvers at risk for total hip dislocation. J. Biomech. 36 (4), 577−591.

Oyama, E., Agah, A., MacDorman, K.F., Maeda, T., Tachi, S., 2001. A modulator neural network architecture for inverse kinematics model learning. Neurocomputing 38, 797−805.

Rabuffetti, M., Baroni, G., 1999. Validation protocol of models for centre of mass estimation. J. Biomech. 32 (6), 609−613.

Robert, T., Cheze, L., Dumas, R., Verriest, J.-P., 2007. Validation of net joint loads calculated by inverse dynamics in case of complex movements: application to balance recovery movements. J. Biomech. 40 (11), 2450−2456.

Saunders, J.B.D.M., Inman, V.T., Eberhart, H.D., 1953. The major determinants in normal and pathological gait. J. Bone Joint Surg. Am. 35-A (3), 543−558.

Xiang, Y., Rahmatalla, S., Arora, J.S., Abdel-Malek, K., 2011. Enhanced optimization-based inverse kinematics methodology considering joint discomfort. Int. J. Human Factors Model. Simulat. 2, 111−126.

Chapter 10

Lu, J., 2011. Isogemetric contact analysis: geometric basis and formulation for frictionless contact. Comput. Methods in Appl. Mech. Eng. 200, 726−741.

Index

Note: Page numbers followed by "*f*" and "*t*" refers to figures and tables, respectively.

A

ADAMS, 90, 111*f*
Ambulation, 149
Analytical IK methods, 44
Animazoo, 209
Anthropometry
  effect of change in, 183
  variations in, 36
Applications, of predictive dynamics, 240–241
  egress/ingress, 242
  ergonomics, 240
  human performance, 241
  injury/disability, simulating, 240–241
  sports biomechanics and kinesiology, 241–243
  testing equipment, digital prototyping, human
    systems integration, 241–242
  unsafe situations, 242–243
Arm lifting motion with load example, 88–90
Arm-leg coupling constraint, 169–170
Asymmetric walking, 184

B

Backward recursive dynamics, 76–77, 84,
    156–157, 189
Basic transformations, 14–16
  definition of, 14*f*
  knee rotation, 16
Benefits of predictive dynamics, 237–240
  cause and effect, 238–239
  Denavit–Hartenberg (DH)
    method, 237–238
  dynamic strength surfaces, use of, 239–240
  effectiveness, 240
  inducing natural behavior, 238
  joint space, use of, 239
  rendering natural motion, 238
  solving dynamics without integration, 238
Box lifting, 121*f*, 197–199, 198*t*, 200*f*, 201, 201*f*
Box-lifting task, validation of, 224–233
  determinants, 225
  discussion, 230–233
  lifting task description, 224
  participants, 225
  qualitative comparison, 225–226
  quantitative comparison, 226–230

C

Cause and effect, 97–98, 176–183, 238–239
Closed-form equations of motion, 86–87
Collision avoidance constraint, 196–197
Composite transformations, 17–19
Configuration, definition of, 10
CONOPT, 57
Constrained joints, 183
Constraints, 46, 103–105, 167–171, 194–197
  collision avoidance, 196–197
  hand distance and orientation, 195–196
  initial and final hand positions, 197
  initial and final static conditions, 197
  vision, 195
  feasible set, 104
  minimal set of, 104–105
  time-dependent, 105–107, 167–170
    arm-leg coupling, 169–170
    dynamic balance, 107
    ground penetration, 106, 168
    joint limits, 105, 167–168
    self-avoidance, 107, 170
    strength limits, 168
    torque limits, 105
  time-independent, 107–108, 170–171
    ground clearance, 108, 170–171
    symmetry conditions, 107–108, 170
    initial and final foot contacting position, 171
Coordinate systems, establishing, 30–35
  9-DOF model of upper limb, 31–32
  DH parameters of lower limb, 32–35
Coriolis torque, 77
Cost function, 46
Coupling constraint, 169–170

D

Data collection, 212–213
Data-driven human motion prediction, 3–4
Degrees of freedom (DOF), 69
  joints as, 151
Delta potential energy, 51–53
Denavit–Hartenberg (DH) method, 25–27, 151,
    237–238
Design variables, 46
Differential algebraic equations (DAEs), 1

Digital human (DH)
  mapping of strength to, 138–140
  modeling, 131–132
Digital human models (DHM), 127
  55-DOF, 38f, 154f, 188f
  muscle-level model of, 128f
  post-processing approach, 140
  pre-processing approach, 140
Digital prototyping, 241
Directed transformation graphs, 19–23
Discomfort joint-limit penalty term, 55f
Discretization and scaling, 108–109
Dynamic balance, 107
Dynamic effort, 101–102
Dynamic equations of motion, 72–74
Dynamic stability, 99–101, 190–191
Dynamic strength surfaces, 239–240
Dynamics sensitivity analysis, 78–80

E

Egress/ingress, 242
Elbow, strength assessment for, 133, 134f
Equations of motion (EoM), 1, 3, 69, 96, 124, 187–190
  closed-form, 86–87
  dynamic, 72–74
Ergonomics, 240
Extended vectors, concept of, 13–14
Eye vector, 195

F

Fatigue, 140–145
  strength and, 145
Feedback to simulation, 233
Foot support polygon (FSP), 107
Foot support polygon, 159f
Force-length relationship, 130–131
Forward dynamics, 97, 97f
Forward kinematics, 24, 24f
Forward recursive kinematics, 76, 83, 155–156, 187–189
Future research, 243–245
  intelligence, 243
  modeling with high level of fidelity, 244
  psychological and physiological factors, 243–244
  real-time simulation, 244–245
  soft-tissue dynamics, 243

G

Gait analysis, 149
Gait model, 158–161
  ground reaction forces (GRF), 159–161

one-step gait model, 158
Gait, predicting, 172–175
  normal walking, 172–175
General optimization problem, 2
General rigid body displacement, 10–12
  rotation and translation, 11–12
Geometric IK methods, 44
Global DOFs and virtual joints, 39–40, 154–155
Ground clearance constraints, 108, 170–171
Ground penetration constraints, 106, 168
Ground reaction forces (GRF), 99–100, 159–161, 164–166, 175, 182, 190–191, 201

H

Hand-orientation constraint, 195–196
Home configuration, establishing, 30
Homogeneous coordinates, concept of, 13–14
Human models, 7
  digital representation of, 8f
  hand position with foot, 8f
Human performance, 241
Human performance measures, development of, 49–57
  delta potential energy, 51–53
  discomfort, 53–55
  effort, 50–51
  joint displacement, 50
  numerical solutions to optimization problems, 57
  single-objective optimization, 55–57
Human skeletal model, 187
Human systems integration, 241–242
Hypertrophied muscles, 129

I

Inertial motion capture systems, 209–210
Injury/disability, simulating, 240–241
Inner optimization, 102–103
Intensity-endurance time curve, 140–142, 144–145, 144f
Interior point (IP) method, 57
Inverse dynamics, 97, 97f, 166
Inverse kinematics (IK), 41–42, 208
  posture prediction versus, 44–45
  analytical and geometric, 44
Isokinetic dynamometer, 132–133
  for elbow strength measurement, 134f
  for knee strength measurement, 133f

J

Jerk, 63, 101
Joint angle limits, 106t, 182t

Joint angle profiles, 88*f*, 90*f*, 179*f*
Joint displacement, 42, 50, 56*f*
Joint limits, 105, 167−168
Joint profiles
    as B-spline curves, 58−59
    discretization, 80−81
Joint space, 127−128
    muscle space versus, 151−152, 239
Joint torque profiles, 88*f*, 91*f*
    for dynamic box-lifting, 201, 202*f*
    with external load, 91*f*
    for stride, 175, 176*f*
Joint-link system, 72, 72*f*
Joint-specific fatigue behavior, 143
Joints as degrees of freedom, 151
Jumping task, 120, 124*f*

K

Kinematic skeleton, 27−30
Kinematic analysis, 7−10
    55-DOF whole body model, 37−39
    215-DOF human model, 37*f*
    anthropometry, variations in, 36
    basic transformations, 14−16
        knee rotation, 16
    composite transformations, 17−19
        example, 17−19
    coordinate systems, establishing, 30−35
        9-DOF model of upper limb, 31−32
        DH parameters of lower limb, 32−35
    Denavit−Hartenberg representation, 25−27
    directed transformation graphs, 19−23
        multiple transformations, 20−23
    extended vectors and homogeneous coordinates,
        13−14
    general rigid body displacement, 10−12
    global DOFs and virtual joints, 39−40
    hand position with foot, 8*f*
    human models, digital representation
        of, 8*f*
    kinematic skeleton, 10*f*, 27−30
    multi-segmental link, position determination of,
        24−25
    rotation and translation, example of, 11−12
    Santos® model, 36
Kinematics sensitivity analysis, 77−78
Kinesiology, 241−243
Knee flexion and extension torque, 135*f*
Knee rotation, 16
Knee strength measurement, 133, 133*f*
Kneeling, 120, 123*f*
KNITRO, 57

L

Ladder climbing, 120, 122*f*
Lagrangian equation, recursive, 75−81
    backward recursive dynamics, 76−77
    dynamics sensitivity analysis, 78−80
    formulation of, 74−75
    forward recursive kinematics, 76
    joint profile discretization, 80−81
    kinematics sensitivity analysis, 77−78
    sensitivity analysis, 75, 77
LANCELOT, 57
Level of fidelity, 27
Lifting, 187
    dynamic stability and ground reaction forces,
        190−191
    equations of motion and sensitivities, 187−190
        backward recursive dynamics, 189
        forward recursive kinematics, 187−189
        sensitivity analysis, 189−190
    formulation, 191−192
    human skeletal model, 187
    input parameters for, 192*f*
    multi-objective optimization, computational
        procedure for, 197−199
        lifting determinants and error quantification,
            198−199
    predictive dynamics optimization formulation,
        192−197
        constraints, 194−197
        design variables and time discretization,
            193−194
        objective functions, 194
    predictive dynamics simulation, 199−201
    validation, 201−204
Lifting task description, 224
Link length and mass properties, 39*t*
Loading effects, 183
Localized muscle fatigue, 140
Lower limb
    4-DOF model of, 33*f*
    DH parameters of, 32−35

M

Marker placement protocol, 210−212, 213*f*
Marker protocol, 210−211
Markerless motion capture systems, 210
Markers, 209−210
Maximum holding time (MHT), 142−143
Mechanical energy, 101
Metabolic energy, 101
Modeling with high level of fidelity, 244
Motion Analysis, 209−210

Motion capture systems, 209–213
  marker placement protocol, 211–212
  optical, 210–211
  overview, 209–210
  subject preparation and data collection,
    212–213
Motion determinants, 209
Motion prediction, 42
Motion prediction formulation, 60
  constraints, 60
  design variables, 60
Motion profiles, 2–3
Multi-body dynamics, 238
Multi-objective optimization, computational
    procedure for, 197–199
  lifting determinants and error quantification,
    198–199
Multi-objective problem statement, 65
  design variables and constraints, 65
Multiple transformations, 20–23
Multi-segmental link, position determination of,
    24–25
Muscle activation-deactivation, 142–143
Muscle hypertrophy, 129
Muscle strength, models of, 127–128, 128f
Muscle versus joint space, 151–152, 239

N

Natural behavior, inducing, 238
Natural motion, rendering, 238
Newton–Euler formulation, 69–70
Normal walking, 172–175
  dynamics, 175
  kinematics, 173–175
Normalization of data, 213–214
Numerical discretization, 171

O

One-DOF elbow joint, 28f
Optical motion capture systems, 209–211
Optimization, 41
  3-DOF arm example, 47–49
  15-DOF motion prediction, 61–65
  constraints, 47
  cost function, 47
  design variables, 46–47
  human performance measures, development of,
    49–57
    delta potential energy, 51–53
    discomfort, 53–55
    effort, 50–51

    joint displacement, 50
    numerical solutions to optimization problems, 57
    single-objective optimization, 55–57
  ingredients of, 46
  joint profiles as B-spline curves, 58–59
  motion between two points, 58
  motion prediction formulation, 60
    constraints, 60
    design variables, 60
  multi-objective problem statement, 65
    design variables and constraints, 65
  optimization algorithm, 62–63
  planning and execution in, 43f
  and posture prediction, 41–43
Optimization formulation
  for lifting, 192–197
    constraints, 194–197
    design variables and time discretization,
      193–194
    objective functions, 194
  for walking, 150, 166–171
    constraints, 167–171
    design variables, 166
    objective function, 166
Optimization problem, in three main
    components, 98f
Optimization-based trajectory generation, 150
Optotrak, 209–210
Orientation constraint, 65, 195–196
Oscillating motion with boundary conditions,
    114–115
  and one state-response constraint, 116–117
  and two state-response constraints, 118–119

P

Posture prediction, 41–43
  3-DOF arm example, 47–49
  15-DOF Denavit—Hartenberg model, 61–62
  15-DOF model, motion prediction of, 63–65
  empirically-based, 44–45
  human performance measures, development of,
    49–57
    delta potential energy, 51–53
    discomfort, 53–55
    effort, 50–51
    joint displacement, 50
    numerical solutions to optimization
      problems, 57
    single-objective optimization, 55–57
  inducing behavior, 43–44
  versus inverse kinematics, 44–45
    analytical and geometric IK models, 44

joint profiles as B-spline curves, 58–59
motion between two points, 58
motion prediction formulation, 60
  constraints, 60
  design variables, 60
multi-objective problem statement, 65
  design variables and constraints, 65
and optimization, 45–47
optimization algorithm, 62–63
task-based approach, 44, 45*f*
Potential energy, 52*f*
Predictive dynamics (PD) approach, 1–2, 95
  constraints, 103–105
    feasible set, 104
    minimal set of constraints, 104–105
    time-dependent constraints, 105–107
    time-independent constraints, 107–108
    types of, 105–108
  discretization and scaling, 108–109
  dynamic stability, 99–101
  example formulations, 120
    for box lifting, 121*f*
    for jumping, 124*f*
    for kneeling, 123*f*
    for ladder climbing, 122*f*
    for running, 123*f*
    for stairs climbing, 121*f*
    for throwing, 122*f*
    for walking, 120*f*
  flow chart of, 98*f*
  inner optimization, 102–103
  performance measures, 101–102
  problem formulation, 95–99
  single pendulum, numerical example of,
    109–119
    oscillating motion with boundary conditions,
      114–115
    problem description, 109–110
    simple swing motion with boundary
      conditions, 111–113
Procedure planning, 43–44

R

Range of motion (ROM) trial, 212–213
Real-time simulation, 244–245
Recursive dynamics, 69–70
  2-DOF arm, using, 81–87
    backward recursive dynamics, 84
    closed-form equations of motion, 86–87
    DH parameters, 82
    forward recursive kinematics, 83
    gradients, 84–85

arm lifting motion with load example, 88–90
dynamic equations of motion, 72–74
general static torque, 70–71
Lagrangian equation, 75–81
  backward recursive dynamics, 76–77
  dynamics sensitivity analysis, 78–80
  formulation of, 74–75
  forward recursive kinematics, 76
  joint profile discretization, 80–81
  kinematics sensitivity analysis, 77–78
  sensitivity analysis, 75, 77
  trajectory planning example, 87–88
Rhomert's curve, 140–142
Rotation matrix, 12
Running, 123*f*

S

Santos® marker protocol, 176–178, 211, 212*f*
Santos® model, 36
Self-avoidance, 107, 170
Sensitivity analysis, 75, 77
  of dynamics, 78–80
  of kinematics, 77–78
  of lifting task, 189–190
  of walking task, 157
Sequential quadratic programming (SQP)
    algorithm, 57, 172, 197–198
Shoulder complex, 28*f*, 29
Shoulder flexion and extension torque, strength
    surfaces for, 136*f*
Sideways and backward walking, 183
Simple swing motion with boundary conditions,
    111–113
Single pendulum, 109–119, 110*f*
  joint angle prediction of, 112*f*, 114*f*, 116*f*, 118*f*
  joint torque prediction of, 113*f*, 117*f*, 119*f*
  joint velocity prediction of, 113*f*, 117*f*, 119*f*
  oscillating motion with boundary conditions,
    114–115
    and one state-response constraint, 116–117
    and two state-response constraints, 118–119
  problem description, 109–110
  simple swing motion with boundary conditions,
    111–113
Single-objective optimization, 55–57
SNOPT (Sparse Nonlinear OPTimizer), 57
Soft-tissue dynamics, 243
Spatial kinematics model, 152–156
  forward recursive kinematics, 155–156
  global DOFs and virtual joints, 154–155
  kinematic 55-DOF human model, 152–153
Sports biomechanics, 241–243

Stability, 102
dynamic. *See* Dynamic stability
Stairs climbing, 120, 121*f*
Static torque, 70−71
Straight knee, 138−139
Strength
  assessment, 132−133
  and fatigue, 145
  influences, 128−132
  joint space, 127−128
  limits, 127, 140, 145, 168
  mapping of, to digital humans, 138−140
  normative strength data, 134−136
  percentiles, representing, 137−138
Symmetric walking, 158
Symmetry conditions, 107−108

T

T-pose, 212, 224
Task planning, 43−44
Tasks of predictive dynamics, 120
  box lifting, 121*f*
  jumping, 124*f*
  kneeling, 123*f*
  ladder climbing, 122*f*
  running, 123*f*
  stairs climbing, 121*f*
  throwing, 122*f*
  walking, 120*f*
Testing equipment, digital prototyping, human
      systems integration, 241−242
Throwing task
  formulation for, 122*f*
  motion capture data for, 213−214
Time-dependent constraints
  for lifting task, 194
  for predictive dynamics, 105−107
    dynamic balance, 107
    ground penetration, 106
    joint limits, 105
    self-avoidance, 107
    torque limits, 105
  for walking optimization, 167−170
    arm-leg coupling, 169−170
    dynamic balance, 168
    ground penetration, 168
    joint limits, 167−168
    self-avoidance, 170
    strength limits, 168
Time-independent constraints
  for lifting task, 194
  for predictive dynamics, 107−108

ground clearance, 108
  symmetry conditions, 107−108
for walking optimization, 170−171
  ground clearance, 170−171
  initial and final foot contacting position, 171
  symmetry conditions, 170
Torque limits, 105
Torque-angle relationship, 131
Trajectory planning example, 87−88
Transformation matrix, 18−19

U

Uneven terrains, walking on, 184
Unsafe situations, 242−243

V

Validation methodologies, 207−209, 214−216
  box-lifting task, 224−233
    determinants, 225
    lifting task description, 224
    participants, 225
    results, 225−233
  difficulties in, 208
  feedback to simulation, 233
  motion capture systems, 209−213
    marker placement protocol, 211−212
    optical, 210−211
    overview, 209−210
    subject preparation and data collection,
        212−213
  motion determinants, 209
  normalizing the data, 213−214
  walking task
    description, 216
    determinants, 217
    participants, 217
    results, 217−224
Vicon System, 209−210, 211*f*
Virtual joints, 39−40, 154−155
Virtual Soldier Research (VSR) program, 1
Vision constraint, 195, 196*f*
Vision vector, 195

W

Walking analysis, 149
Walking, biomechanics of, 149−151
  asymmetric walking, 184
  cause and effect, 176−183
  different terrain types, walking on, 184
  dynamics formulation, 156−157
    backward recursive dynamics, 156−157

mass and inertia property, 157
   sensitivity analysis, 157
effect of changing anthropometry, 183
effect of changing loads, 183
effect of constrained joints, 183
formulation for, 120*f*
gait model, 158–161
   ground reaction forces (GRF), 159–161
   one-step gait model, 158
gait, predicting, 172–175
   normal walking, 172–175
ground reaction forces (GRF), 164–166
joints as degrees of freedom, 151
muscle versus joint space, 151–152
numerical discretization, 171
optimization formulation, 166–171
   constraints, 167–171
   design variables, 166
   objective function, 166
sideways and backward walking, 183
spatial kinematics model, 152–156
   forward recursive kinematics, 155–156
   global DOFs and virtual joints, 154–155
   kinematic 55-DOF human model, 152–153
uneven terrains, walking on, 184

Zero-Moment point (ZMP), 161–164
   calculation, 163–164
   global forces at origin, 163
   global forces at the pelvis, 162
Walking determinants (WD), 217
Walking task, validation of
   description, 216
   determinants, 217
   discussion, 221–224
   participants, 217
   qualitative comparison, 217–220
   quantitative comparison, 221
Working, of predictive dynamics, 2–3

X

Xsens, 209

Z

Zero-moment point (ZMP), 99, 150, 161–164
   active forces, 99–100
   calculation of, 100, 163–164
   global forces at origin, 163
   global forces at pelvis, 162
   passive forces, 99–100

Printed in the United States
By Bookmasters